GUIDE TO THE

Design, Selection, and Application of Screw Feeders

GUIDE TO THE

Design, Selection, and Application of Screw Feeders

Lyn Bates

Professional Engineering Publishing

Professional Engineering Publishing Limited
London and Bury St Edmunds, UK

First Published 2000

ISBN 1 86058 285 0

A CIP catalogue record for this book is available from the British Library.

About the Author

The British Materials Handling Board, following their long-term promotion of bulk storage and handling interests, perceived the need to make available a practical guide to the design, selection, and application of screw feeders. The author, Lyn Bates, as an international renowned expert in this field, was commissioned to prepare this user Guide.

As Managing Director of Ajax Equipment Limited, a company that specializes in screw-type equipment for solids handling, he enjoys a 'hands-on' attitude to powder handling problems. He has introduced various design innovations and patents in the field and designed various instruments for measuring flow-related powder properties. As a member and past chairman of the Institution of Mechanical Engineers Bulk Materials Handling Committee, he produced a 'Guide to the Specification of Bulk Solids for Storage and Handling Applications'. An active member of the European Federation of Chemical Engineers Working Party on the Mechanics of Particulate Solids, and sitting on various BSI and other technical committees, he is dedicated to promoting education in this specialized section of engineering.

Lyn Bates has written many technical papers and publications on aspects of bulk solids handling. This book complements related publications by the BMHB, which include the author's earlier book 'A User Guide to Segregation', as well as 'User Guide to Particle Attrition in Mechanical Handling Equipment', prepared by a working party chaired by Lyn Bates.

RELATED TITLES

Title	Author	ISBN
An Introductory Guide to Flow Measurement	Roger Baker	0 85298 670 X
An Introductory Guide to Industrial Flow	Roger Baker	0 85298 983 0
Process Pump Selection – A Systems Approach	John Davidson Otto von Bertele	0 85298 670 X
Hopper and Silo Discharge – Successful Solutions	IMechE Seminar Proceedings	1 86058 192 7
Successful Pneumatic Conveying	IMechE Seminar Proceedings	1 86058 195 1
Powder to Bulk	IMechE Conference Proceedings	1 86058 272 9
Bulk Materials Handling: Real-Time Analysis	IMechE Seminar Proceedings	1 86058 106 4
IMechE Engineers' Data Book – Second Edition	Clifford Matthews	1 86058 248 6
A Guide to Presenting Technical Information – Effective Graphic Communication	Clifford Matthews	1 86058 249 4
Handbook of Mechanical Works Inspection – A guide to effective practice	Clifford Matthews	1 86058 047 5

Professional Engineering Publishing (publishers to the Institution of Mechanical Engineers) offers a valuable and extensive range of engineering books, conference and seminar transactions, and prestigious learned journals, as well as a successful list of magazines. For the full range of titles published, contact:

Sales Department
Professional Engineering Publishing Limited
Northgate Avenue
Bury St Edmunds
Suffolk
IP32 6BW
UK
Tel: +44 (0) 1284 724384
Fax: +44 (0) 1284 718692
E-mail: sales@imeche.org.uk
www: pepublishing.com

Contents

Chapter 1

Introduction

Manufacturing industry is the foundation of universal prosperity. More than 60 percent of all products consumed and handled by man are at some time in the form of bulk solids, many of which pass through several handling and processing operations. Intermediate storage and controlled rate discharge figure largely in these production requirements. Increasingly, the need for reliable and predictable performance is paramount to the efficiency and quality of manufacturing processes and plant performance. The Rand report indicated that, in general, plants handling loose solids have a far inferior performance to those that handle liquids and gases. It went on to say that, despite the advances made in powder technology, the situation had not significantly improved from plants commissioned in the 1960s. The root of most problems encountered was not due to any failings in the process technology, or of the basic engineering construction, but invariably lay in the failure to accommodate behavioural properties of the bulk material handled.

Clearly, an undervaluation of education in this field is impeding progress towards radical improvements in industrial performance. An important aspect of securing improvement in this field is achieving an understanding of bulk material behaviour, and its significance in the selection and specification of storage and feeding equipment for loose solids. One common difficulty is identifying relevant properties of solids for contractual purposes. This situation was addressed by the Bulk Materials Handling Committee of the Institution of Mechanical Engineers, who prepared a *Guide to the Specification of Bulk Materials for Storage and Handling Applications.* Also the British Materials Handling Board published a *Guide to Powder Testing*, a *Guide to Particle Attrition in Solids Handling Equipment* and a *Guide to Segregation*, to provide background on some common phenomena affecting the performance of solids handling plant.

Recognizing the need for practical advice on the choice and specification of solids handling equipment, this book has been constructed to guide potential users through the criteria for selecting and specifying screw feeders. It also highlights technical and economic factors relating to the design and use of screw feeders.

The history of screw equipment is steeped in antiquity. The first recorded use of screws for materials handling is attributed to Archimedes (287–212 BC) who designed screws to elevate water from the holds of ships for King Heiro of Syracuse. Similar devices have since been extensively employed for irrigation, operated manually, by animals, wind, and more recently by internal combustion motors and electric power. Some modern units used for elevating fresh and sea water, as well as fluids such as raw sewage, attain dimensions exceeding 2 m in diameter.

The use of screw equipment for handling bulk solids is more recent. The first mechanized application of helical screw devices for conveying powdered materials is credited to an American engineer named Evans, who installed screw-type devices for transporting flour in a grain mill built in 1785. Those machines used wooden paddles arranged to form a helical surface around metal shafts. Wider use of screw flights made by pressing metal ribbon-shaped discs into screw segments inspired Frank C. Caldwell to patent a flight-forming machine to make continuous runs of screw flighting from metal strips. Both methods of manufacture are still in use. Special sections of material are employed for continuous spiral forms, from round, square, strips, and triangular sections. Latest technology in laser cutting allows complex profiles to be cut for ribbon-form flighting and differing geometrical forms used for mixing duties, and the like.

Screws gained extensive use in agriculture for grain handling, both as separate units and as integral parts of equipment, as in combine harvesters.

The introduction and development of mass production techniques directed interest to automated handling. The simplicity, enclosure, and compactness of screw conveyor based handling encouraged wider industrial applications, under the pressures of manufacturing scale and the economics of reducing manual labour. Extensive use was made of screws for simple conveying duties, particularly for 'bridging' between different stages of continuous process operations. Feeding and elevating applications involve more technical factors, relating to the interaction of screw mechanics with the complexities of bulk material behaviour, hence suppliers in this field tend to be specialized, although many standard designs are adopted for repetitive and simple duties.

Screw feeders today play an increasingly important role in the drive towards improved quality, reduced costs, increased capacity, better working conditions, and flexibility in solids processing. Advances in control methods are being matched by improved predictability and reliability of the processes being controlled. The intensive and integrated nature of many production lines crucially depends upon each element working to its full design capability. Solids feeding operations comprise a key activity, renown for operating difficulties out of all proportion to the cost of the equipment.

Reasons driving the growth in use of gravimetric feeders include the need for verification of performance, the provision for alarm or fail-safe action in cases of failure, and accompanying improvements in the accuracy of the metering process itself. The standard of accuracy has improved in recent years, and the continuous erosion of 'give-away' or 'over-delivery' of product to filling machines has resulted in impressive savings. Similarly, the quality and consistency of a product, whether pharmaceutical, food, chemical, or whatever, is dependent upon a close control of the ingredient materials.

Many manufacturers now serve the market. Some of these offer standard units for well-proven applications, or for the user to determine their suitability. Others design for use, based upon experience and good practice. While many similar types of applications recur, permutations of the range of bulk products with differing applications and environmental conditions are so vast that it is common to find new duties for which no prior identical example of use can be drawn upon for performance verification. In such circumstances, the application criteria and basis of specification need to be established with some precision, as no feeder, screw type machine, or indeed any other form of handling equipment, can guarantee to handle a bulk material of unspecified condition.

It was not until Jenike developed a theory of flow in hoppers, in the late 1960s, that a sufficient understanding of the mechanics of solids was gained to facilitate a more scientific basis for the design of screw feeders and their hoppers. Development since has advanced by leaps and bounds, both with regard to innovative designs of hopper geometry, and the exploitation of variants in screw design. Research in the field is inevitably concentrated upon specific and relatively narrow technical aspects of particle and bulk technology, whereas many developments in screw feeders, their supply hopper geometry, and specialized features and accessories, are application driven. It is also the case that many feeder designs are manufactured for specific duties, and are never included in a published catalogue. For this reason, a guide based upon practical design and usage offers the means to bring together a summary of the current state of the art, to aid the non-specialist in the selection and specification of screw feeders for a wide range of duties.

1.1 Screw applications

Arising from the ability to move loose bulk solids along the axis of a confined helical blade, screw equipment has been widely adopted by industry for a great variety of solids handling duties, ranging from ship unloaders and high-capacity conveyors, to dispensing devices that meter small quantities of powder. Use is also made of this form of equipment in innumerable process applications, such as heat transfer duties and both high- and low-temperature conveying. Mixing and blending is also carried out in differing ways by helical screws and variants, such as ribbon constructions, discontinuous flights, crescent and paddle blades, and a host of other shapes.

The flexibility of screw handling is also exploited in compacting devices, de-watering screws, packing and filling machines, crammer screws to feed extruders and roll presses, and for cookers, blanchers, driers, and similar functions that require the movement of loose solids in a continuous stream.

Early industrial use of screws centred on repetitive handling duties, as with grain and flour handling. The urge to save manual labour and deal with higher quantities of material prompted wider uses and many mis-applications. However, as an understanding grew of their advantages and limitations, the ingenuity of engineers produced a plethora of applications in all industries, from food and pharmaceuticals to waste and sewage handling. There is now no section of industry dealing with bulk material that does not employ some form of helical screw-type equipment.

Screws have found ubiquitous usage because they offer the following favourable features:

1. They have a basic simplicity of design, construction, and operation, have few moving parts, require relatively imprecise fabrication limits, and offer robust construction and reliable performance. Fabrication can be in mild or stainless steels, abrasive resisting materials, or plastics, with finishes for hygiene, and corrosion and wear resistance.

2. The cross-section of the machine is compact. The flow-promoting component has a single run. (No return path has to be accommodated as with a belt or chain conveyor.)

3. Enclosure can provide for safety, weather protection and washing down, and the containment of dust, gases, vapours, internal or external pressures, and even explosion containment if required.

4. The equipment can be designed to stop and restart under load, accommodate a controlled or flood feed at the inlet, be reversible, and accept multiple inlets and outlets. Extended inlets may be provided to collect from long slots, which can be either of a flood feed nature with live extraction along the whole length, or of a type which accepts an erratic in-feed and is only required to clear eventually.

Note that some of the above facilities are not mutually compatible.

5. The equipment can be made in a wide range of sizes, from about 10 mm diameter to in excess of 2 m diameter. Operating speeds are effective from a fraction of a r/min to over 500 r/min, although most duties employ screw speeds in the range of 20–100 r/min. This variable capacity is used to control transfer rates; hence many metering duties are undertaken with helical screw devices.

6. Multiple screws, variable geometry, inclination from horizontal to vertical, jacketed casings, 'shaftless' and ribbon screws, 'plug seals', and a host of design variants offer scope for innovative and specialized functions.

There are also a number of limitations and disadvantages of helical screws that can inhibit the application of screws, when compared with other forms of bulk handling equipment.

1. The interaction between the screw and the media handled introduces a behaviour relationship, which must be satisfied for effective operation. The mechanical efficiency of transport is low in comparison to belt conveyors, where the bulk material is less disturbed in transit.

2. Screw equipment is not entirely self-cleaning of the product conveyed, because there is an essential operational clearance between the screw flight and the wall of the casing in which it rotates. Various design techniques are employed to facilitate cleaning, as may be required. The flight tip clearance is also a potential hazard for trapping and fracturing granules that wedge in this clearance space.

3. Whereas simple applications, such as screw conveying, can generally be reliably sized and assessed for power requirements, many forms of screw feeder, elevators, specialized and process-type screw equipment, require specialized knowledge or representative tests in order to prove their performance. Many mechanical designers have limited knowledge of bulk material flow properties, and limited access to powder testing equipment; hence some forms of screw equipment remain the domain of specialists. The complexity of bulk material behaviour within the working region of the screw inhibits an understanding of the overall performance characteristics of the equipment as an integral unit.

Within this balance between the advantages and limitations of screw equipment, empirical developments and advances in the technology are expanding the frontiers of applications. Standard forms of equipment are available, sometimes from stock. Custom designs, to serve specific duties, take two general forms. Standard components and assemblies are commonly offered as 'off-the-shelf' composites, to suit particular dimensions and capacities of simple applications by incorporating choices of differing standard parts. By contrast, custom-built 'specials' are designed and manufactured as 'one-off' units to suit specific dimensions and requirements, often incorporating design features particular to the application.

The need to control the feed rate of bulk materials is very common throughout industry. The use of screw feeders to dispense powders, granules, pastes, and other particulate forms of products is prolific. Foods, pharmaceuticals, plastics, pigments, minerals, chemicals, sewage, and a host of diverse industries and products are served by screw feeding equipment. While most applications work at ambient temperature and pressure, some feeders operate at temperatures from cryogenic levels up to 1000°C, and at pressures from vacuum to those commensurate with vessels for conventional duties in normal and varied chemical atmospheres.

In general, feed screws are used to handle fine powders or small particulate products but, with special design consideration, exceptional duties such as the feeding of house bricks, metal punchings, and fibrous products have been undertaken. The suitability of feed screws is more usually dictated by the need to secure reliable flow from the supply hopper, rather than by any problem of the screws moving the product.

Feeder applications range from less than 1 kg/h to over 100 tonne/h, delivered by means of screws from 10 to 600 mm diameter. It is unusual for single span screws to be more than 8000 mm long, because excess deflection allows the screw to rub on the casing. The section of screw exposed to a flooded hopper is rarely longer than 4000 mm in the case of non-mass flow applications, or 2000 mm to serve mass flow extraction duties. Screws are used singly, as twins, or in multiple arrays. The usual reason for using more than one screw in a feeder is to secure a wide opening for reliable flow.

Feed screw rotational speeds may be fixed or variable, according to the type of discharge control required. Typical working speeds are in the range of 15–100 r/min. Within these speeds the output volume varies linearly with speed, in fact this direct relationship of feed rate with output holds down to extremely low speeds of screw rotation. The feature which most affects feed regularity at very low screw speeds, is how the material falls from the end of the screw. At high rotational speeds the ability of the material to fill the screw volume at a stable density is impaired by the dilatation of the bulk material moving at high flow rates, and the manner in which the material can attain the velocity to fill behind the moving parts.

Screw feeders are used as independent units, or in combination, for many process operations. They are also used as integral components of equipment, such as roller presses and pin mills, and other powder processing machines. Their ubiquitous use is an essential feature of modern solids handling plant. Over this wide range of duties, users require a basic understanding of key performance-related features, in order to select equipment providing the best performance.

1.2 Properties of bulk solids

Particulate solids may be considered as a fourth state of matter, incorporating all the complexities of solid mechanics, chemistry, physics, electrostatics, and multi-phase rheology for any conceivable flow situation. Some behavioural circumstances are quite easy to understand, or are at least predictable. Unfortunately, like people, few bulk solids are normal, uniform, consistent, and stable under all circumstances; hence they may be viewed as having personality characteristics with their behaviour susceptible to environmental conditions and prior experiences.

Table 1.1 The personality of bulk solids

'Normal' *Stable and consistent behaviour*	**'Neurotic'** *Awkward to handle*
Very predictable, even, and uniform nature Insensitive to handling and environment Invariable with time and place Easy to accommodate and control	Behaviour varies widely with circumstances Unstable in state and condition Sensitive to surroundings and handling Prone to run amok or lose mobility
'Schizophrenic' *Behaviour changes completely*	**'Masochistic'** *Suffers readily*
Absorbs liquor, hardens, or dries out Sensitive to heat, segregates easily Forms strong, unwelcome bonds	Fragile, delicate, easily breaks down Degrades, sensitive to contamination Deteriorates rapidly, has to be handled with care
'Paranoid' *Hypersensitive to own condition*	**'Sadistic'** *Aggressive to surroundings*
Sensitive to size, shape, appearance, purity Alarmed by unquantifiable concerns Plagued with constraints or quality considerations Obsessed by regulations, bounds of acceptance	Abrasive, toxic, hot, or inflames readily Unpleasant to touch or be near, irritates Fouls or contaminates the locality, corrosive
	'Plain nasty' *Obstinate, dangerous, hazardous*

The performance of a screw feeder is dependent upon the physical properties of the bulk material to be handled in a variety of circumstances. As the output is basically volumetric the mass rate of discharge is directly related to the bulk density of the material. This is never a single value parameter for a bulk material because, even if the particles of composition themselves have a firm and stable structure, the manner in which they nest together can be highly variable. The effective bulk density in which the material occupies the screw volume is, therefore, an important measured value. Screw equipment has to interact with the material to initiate and sustain bulk movement.

The conditions for commencing flow in a confined situation are completely different from those at which flow then continues. Power requirements are also sensitive to apparently minor features of the design. The specification of equipment and the selection of a suitable drive unit must, therefore, be carefully matched. Only in the most straightforward applications, such as screw conveyors, are well-proven formulas published relative to a wide range of duties and different bulk products handled.

1.2.1 Wall friction

Wall friction is an important parameter for all types of screw equipment. Efficient progress along the axis of the machine requires the bulk material to slide on the face of the screw blade. The ease with which this takes place influences the power needs of the unit. It also determines the material movement in an inclined screw. Friction is influential in the design of the storage container for the feeder. Friction on the vertical walls restrains forces acting to compact material in the lower regions and the inclination of converging faces, for both mass flow designs and for self-clearing of the contents, are dependent upon wall friction.

Placing a sample of material on a surface, and then inclining the face until the material slips down the slope, can crudely measure wall friction. However, this test does not distinguish between static and dynamic friction, nor does it bring out any tendency for the material to cohere to the surface. A more rigorous technique is to measure the resistance to sliding of a sample pressed against the contact surface over a range of differing pressures, see Fig. 1.1. A graph of the results in Fig. 1.2 shows the angle of friction and how this is affected by contact pressure. The presence of an intercept of the angle with the value of zero contact force indicates any cohesive forces tending to make the powder adhere to the surface.

1.2.2 Shear strength

Shear strength of a bulk material in differing states of dilation is a key property of interest for flow considerations. The conventional hopper design method for mass flow is based upon 'critical state' theory, and a Jenike shear cell is used to secure yield locus values upon which a design procedure is based. This technique is universally accepted, but not widely used for small hoppers for various reasons. Significant cost and expertise is required to obtain accurate values, compared with full-scale trials and

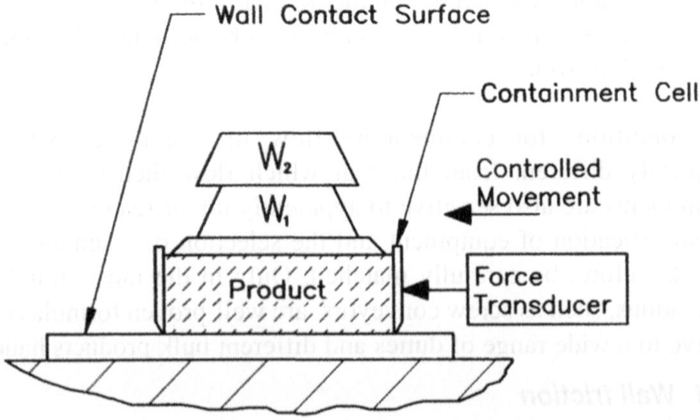

Fig. 1.1 Wall friction measurement

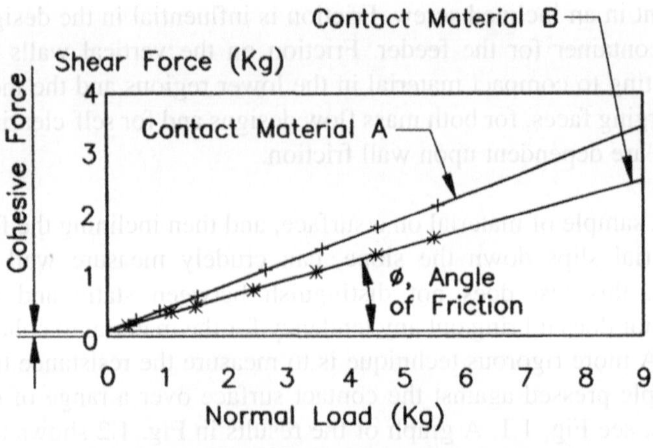

Fig. 1.2 Wall friction angle and cohesion

experience. In small-scale equipment, flow forces are sensitively small, with little scope for allowances for operational uncertainty, safety margins, or conservative factors. However, the hopper interface geometry given by a feed screw offers optimum flow characteristics. The Jenike test procedure is described in the Institution of Chemical Engineers' publication, *Standard Shear Testing Technique,* and a reference test material, CRM 116, is available from the Community Bureau of Reference, for user verification tests. Apart from large-scale applications, where the cost of non-performance or retrofit is prohibitive, there are many crucial feeding duties where the need to attain first time reliable behaviour justifies a detailed investigation of this design, regardless of the capital cost of the equipment. In all cases, it is sound policy to review the potential liabilities of a feeder failure in a risk assessment, to establish what degree of investigation costs is justified. Such an evaluation also allows the significance of any differences in the capital cost of alternative equipment to be placed in a realistic perspective.

More important than sustained flow behaviour, in many instances, is the static strength of the bulk solid. This is relevant to the initial shear of the material, to allow the feed screw to rotate, and also to ensure that further material will collapse from the hopper outlet in order to continue the supply. Starting loads of a screw in a flood-feed mode are invariably more severe than maintaining the drive when the material has attained a flowing condition. To measure the initial shear value it is first necessary to establish the relevant condition of the sample to be tested. Unlike liquids and continuous solids, the strength of a particulate mass is highly variable. The initial failure strength of a bulk mass depends upon the stresses currently acting upon the material and the specific closeness of the particle packing, as determined by the stress history of its bulk formation. The particular density of the bulk gives a useful measure of how closely the particles are packed together, and therefore serves as a measure of its potential condition. In conventional shear testing for flow, samples are compacted at one value of stress and sheared under lower stresses acting normal to the shear plane, to replicate flow conditions where shear is sustained without change in density, see Fig. 1.3. By contrast, incipient failure testing measures the shear strength of the bulk in the presence of the formation stress, to represent confined shear of fine materials. This initial shear value is relevant to the starting load on a feeder screw handling powders in flood-feed condition.

In order to measure bulk strength in the absence of confinement, as relevant to the stress conditions on the underside of an arch, a failure test is carried out after the formation stress is removed. This test reflects the failure conditions on the surface of an arch subjected to passive wall pressures generated by a mass flow hopper, and is measured by an 'unconfined failure' test, as shown in Fig. 1.4. A column of material is compacted in a cylindrical cell and then subjected to axial loading after removal of the cell walls. This is a delicate operation, due to the sensitive nature of the samples and the effect of wall friction opposing the compacting forces. Frictional effects rapidly magnify with the length of

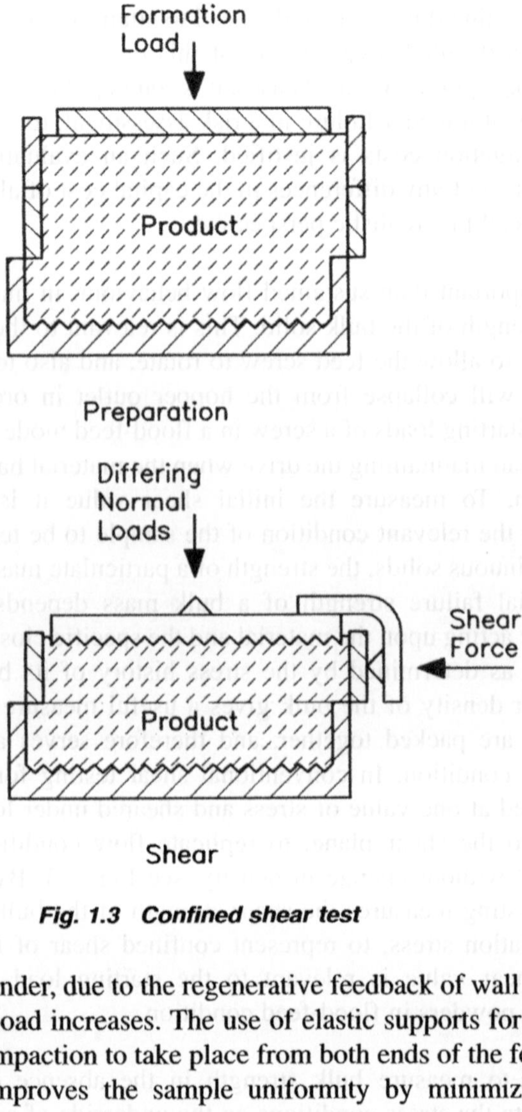

Fig. 1.3 Confined shear test

the compaction cylinder, due to the regenerative feedback of wall pressure as the compacting load increases. The use of elastic supports for the cell, which allow the compaction to take place from both ends of the formation cylinder, greatly improves the sample uniformity by minimizing wall friction effects.

A different approach is used to reflect the shearing of end supports for an arch over a non-mass flow hopper outlet. In this case, the principal stress causing the arch to fail is generated by the weight of product supported over the opening. For this purpose a vertical shear-type test is conducted, see Fig. 1.5. For all such tests, the condition of the sample must reflect the loading conditions experienced by the material in the situation under consideration. Many bulk materials exhibit long-term variations of condition, and may be sensitive to changes in the local environment. Chemical, thermal, or bacterial changes which affect the way that a bulk material behaves must be taken into account when assessing the design or suitability of a feeder. Bulk materials that sinter, gel, or cake into a virtually solid mass should not be expected to flow or shear in a feeder. The circumstances which allow this condition to develop must therefore be avoided, either by treatment, reducing the plant life of the material to safe periods, avoiding the stresses, temperatures, and moisture conditions that cause the problem, or by other methods that are appropriate to the condition in question.

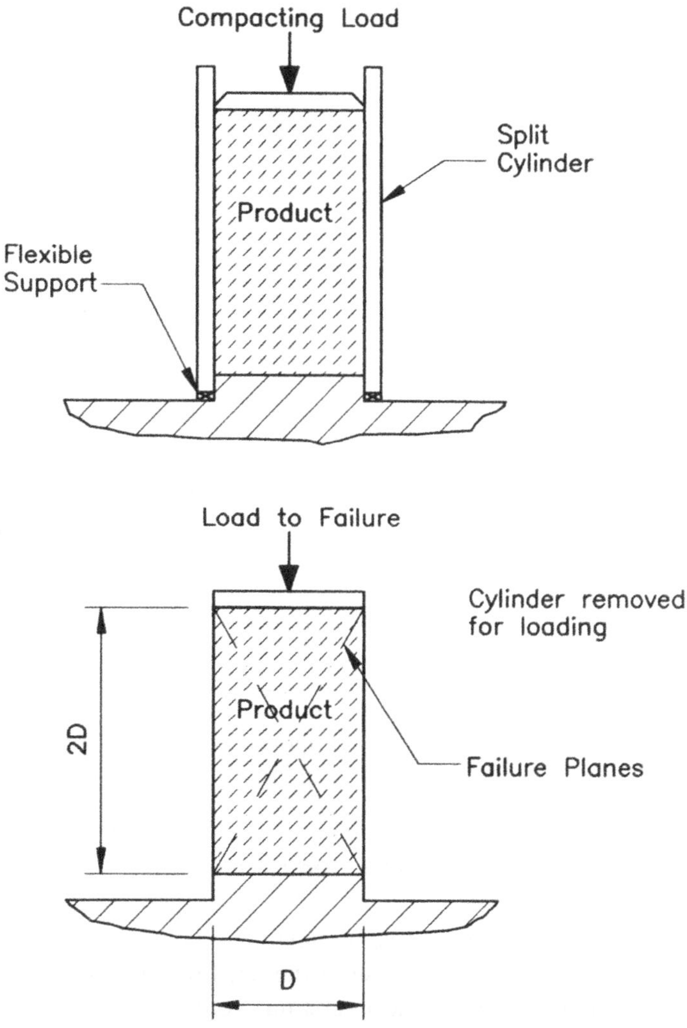

Fig. 1.4 Unconfined failure test

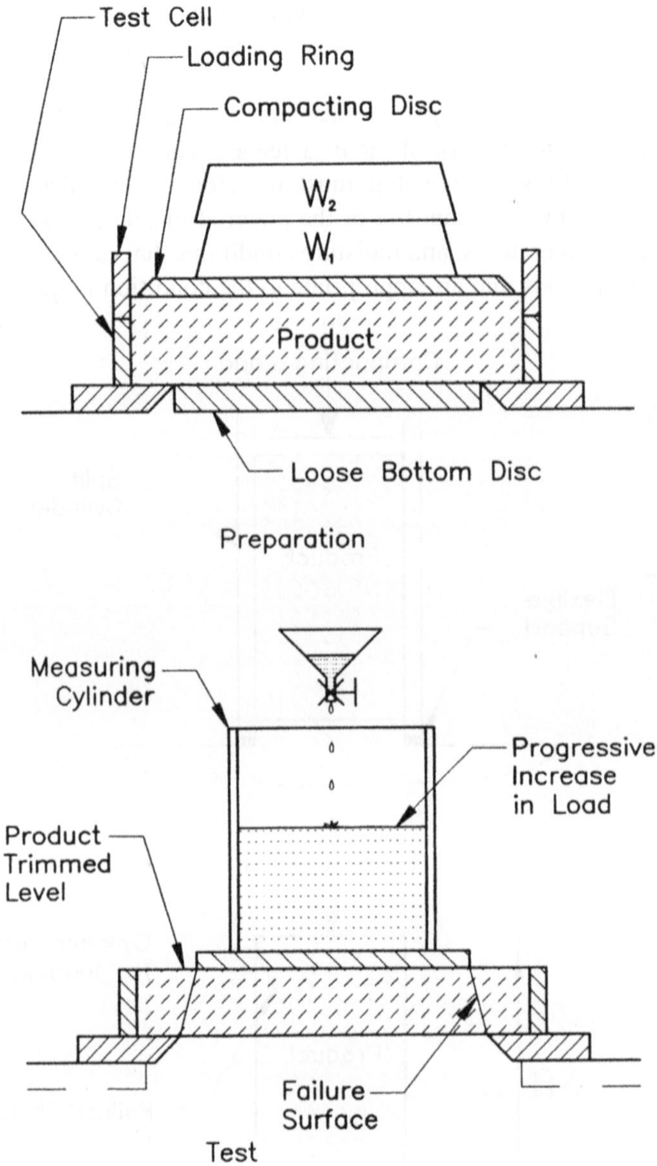

Fig.1.5 Vertical shear test

A common problem with fine particle material is that of changing density for flow or settlement, because of the need to ingest or express gas from the changing volume of the interstitial voids. As the particles assume a close proximity, the reducing permeability of the mass increasingly opposes this second phase flow of interstitial gas. Consequently, fine powders can remain in a fluid condition for extended periods before settling to a stable state, in which condition they then exhibit poor flow characteristics. The only escape route for the gas is towards an unconfined surface; therefore, the path along which the gas has to travel can be long and tortuous, depending upon the size and shape of the storage container.

This situation is exacerbated at elevated temperatures, when the gas viscosity is increased, so material from dryers and kilns can be much more fluid in condition, and for longer, than at room temperature. In order to achieve a reliable performance, a feeder must operate with material in a consistent and stable flow condition. For such materials a mass flow type of hopper is essential to ensure the material has time to settle after loading, and does not follow a preferential flow route through the bed of stored material. Even so, it may be necessary to pay careful attention to the filling process, the size and design of the hopper, and possibly take special steps to secure suitable 'state' control of the powder, to secure reliable results.

Large-grained particles tend to settle quickly to a stable condition of consistent density. Even so, the density they achieve varies according to the method of formation. The most regular of particle shapes, ball bearings, display a wide range of stable packing structures, and then they will change if vibration is applied to the bed. For a shear plane to develop, the particles have to move past each other, and this requires a degree of local expansion in a settled array. When such a packed array is confined by local walls and by a deep bed of stored product, the only source of expansion is at the expense of compaction in another region of the local bulk. The very nature of a bed of hard grains is that it will very strongly resist compaction; as a consequence the initial shear strength of the material can be exceptionally high. Crystalline products, such as salt, granular sugar, and like materials, therefore present considerable difficulties for starting loads on the drives of large-scale screw feeders, unless provision is made for relieving these shear plane expansion demands.

The effect of free surface liquor content in a bulk material, as opposed to bound moisture or water of composition, is to alter the physical and geometric structure of the mass. Water is a common component of liquor/solids mixtures and can be present in a wide range of proportions, categorized by descriptions that reflect increasing moisture content as:

- dry;
- damp powders;
- wet powders;
- 'cake';
- unsaturated paste;
- saturated paste;
- sludge;
- low-concentration slurry;
- high-concentration slurry;
- dispersion.

The range of 'solids handling' relates to conditions up to a saturated paste state. That is where the voids are completely filled with liquid and the material is not compressible, but significant particle contacts provide a shear strength to the compound. Below this condition of liquor content a portion of the voidage space is occupied by ambient gas, and the material may be compacted by the application of vibration and/or the application of external stresses. Where the moisture content is sufficient to occupy all the void space for a material in a strongly compacted condition, but the material is actually at a state of lower density because of air trapped in the voids, the stresses within the bulk are mainly taken by particle-to-particle contact.

There is a sensitive range of moisture content over which the material can change from being unsaturated to a fully saturated state, without variation of moisture content, depending upon the closeness of packing of the particles. For example, the void volume between uniform diameter spherical particles varies from about 30 percent as a close-packed hexagonal array to 40 percent in a random order of packing. A loosely packed material with a moisture content above the critical void filling value will suffer liquefaction when subjected to vibration or sustained shear 'working', as during transport by ship or screw conveyor. This accounts for materials such as filter cakes and centrifuged material changing from a friable wet bulk to an amorphous, plastic, clay-like product during movement along a conveyor, and leads the instability of ships carrying wet coal or ores in heavy seas. The condition is irreversible because once the liquor occupies all the void space it is not possible for the particles to separate uniformly to re-admit air.

The immediate physical effects of moisture are most pronounced with fine particulate materials, because these are more variable in density condition and offer wider shear strength changes than coarse particulates. At low concentrations a moisture film on the surface of particles introduces additional attractive forces in the form of surface tension, as minute cusps form at the contact points. As the moisture content rises, the growth of larger cusps and formation of local regions of saturation leads to a reduction in strength. For a given material there is a moisture proportion that creates a maximum bulk strength condition. For some coals this is in the region of 12 percent moisture.

The highest bulk strength condition occurs when the moisture is not quite sufficient to fill the total void space. The product then continues to gain strength with compaction, leading to poor flow properties in feeder hoppers and high powers being required to shear the material and scrape over confined beds of product, as in the boundary layers of a screw feeder casing around the screw flight clearance.

The flow and compaction nature of fine powders is therefore sensitively influenced by even minor variations of moisture content. Fine products that are hygroscopic can present special difficulties in their handling, as can condensation in equipment prone to present contact surfaces that fall below dew point in temperature.

Another form of problem arises with materials that are soluble or form crystal bridges between particles in the presence of moisture. Products such as salts and sugar will 'cake' due to the formation of multiple binding attachments between the particles in contact. Serious 'caking' achieves bulk strength conditions well beyond the ability of flow channels to generate gravity flow, and often exceeds the power available in mechanical equipment to overcome the set condition. In screw-type equipment the formation of a 'caked' layer in the clearance space between the screw and the casing of the conveyor or feeder can offer a high tip-binding force. This force may lead to wear or power problems, even though the bulk of material in transit is in a loose state.

Moisture also influences the surface contact characteristics of the bulk material. At low concentrations it tends to cause materials to stick to surfaces, but at higher values, or higher contact pressures, it will lubricate the surface and allow the material to slide more easily. The nature of the contact surface is important, as hydrophobic materials, such as high-molecular-density polyethylene, will tend to reject moisture and therefore encourage the slip of wet materials in chutes and hoppers.

Products that are chemically active, such as ground phosphates, or thermally sensitive, such as plastic granules and powder, can also form strong particle-binding forces. For each product under consideration the uniformity, consistency, and stability of composition, throughout its total plant life of process and ambient variation, must be related to the prospect of any physical feature of the material that is able to influence the strength of the particulate mass.

The field of particulate solids is challenging. Apart from the problems associated with measuring behaviour-related properties, the nature of the material itself first has to be closely defined. Table 1.2 lists some of the impediments to registering definitive properties of a bulk product. This provides a clear indication of the need to secure representative samples and measure specific values, rather than expect to secure valid design figures from a database. The authentication of samples and their bounds of variation for a particular project of application is a crucial feature when contractual performance guarantees are involved. It is good practice for their measured values to be closely defined and recorded as key design data.

Table 1.2 Problems of specifying bulk solids condition for handling duties

- There is a prodigious variety of products, constantly increasing
- Most have enormous variations of condition, e.g. particle shape and size distributions
- Products are not often uniform, consistent, and /or stable
- Properties include mechanical, physical, chemical, thermal, and electrostatic
- These properties interact with process conditions
- Some products are sensitive to isotropy, strain rate, or stress thresholds
- Behaviour is often dependent upon the product state and stress history
- Features of interest vary widely; only some can be quantified
- Many are not user-friendly, with relation to human contact and/or equipment
- Particulates have at least two phases, and often three: solids, gas, and liquid
- The material may change during, or as a result of, the handling process
- The extent of variations may not initially be known
- Applications vary enormously, both in duty and in scale
- Permutations of environment with time and process changes are multitudinous
- Ambient conditions vary widely, and sometimes unpredictably
- All these variations compound, rather than act in isolation
- In some cases, material does not exist in sufficient quantities to test, if at all
- Testing resources and application experience tend to be limited, and not inexpensive
- Products rarely travel well for testing appropriate site conditions

Chapter 2

Classes of Screw Equipment

Helical screws are used for moving bulk materials in a number of different ways. While the fundamental operating principle on which screws transport loose solids is based upon the rotation of an inclined face to promote the bulk material to move, screws work under differing loading conditions, at differing inclinations, and with many variations of functions. These differing screw types may be broadly classified as conveyors, elevators, and feeders. The class of feeders includes the use of discharge screws as an integral feature of a bulk storage facility, and as dispensing devices, where the prime function is to meter the feed at a controlled rate. The main distinction between the three main classes of screws for handling solids is based upon the mode of conveying of the material in transit, see Table 2.1.

The selection criteria for these various modes of conveying are completely different. The capacity, power, and interfacing needs occupy distinctively separate considerations. For example, the nature of the material handled and the need to minimize wear may compromise the speed of operation of conveyors. The size of an elevator may be determined more by the casing span than the handling capacity, and the design of a feeder is affected by both the arching potential of a powder and the extraction pattern that it is necessary to generate.

Table 2.1 The three main classes of screw for handling solids

Screw type	*Mode of conveying*	*Description*
Conveyors	Gravity mode	Where the material slides down the face of the screw flight as an inclined plane, as in Fig. 2.1.
Elevators	Dynamic mode	Where the material in transit is rotated to form a continuous annular vortex, only restrained by boundary friction on the casing, as in Fig. 2.2.
Feeders	Flooded mode	Where material occupies the full cross-section of the screw and is promoted to move by the rotating face of the screw blade acting as a moving inclined wedge, as in Fig. 2.3.

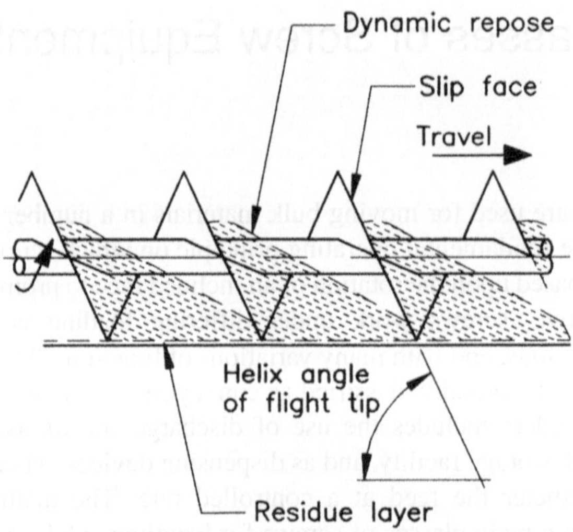

Fig. 2.1 Gravity mode of conveying

2.1 Screw conveyors

Gravity mode conveying is the manner in which conventional screw conveyors operate. These normally work horizontally or gently inclined, with a cross-sectional loading up to around 45 percent fill. If the fill level exceeds the height of the centre shaft, material is carried over into the preceding pitch space and not moved forward (Fig. 2.4). The cross-

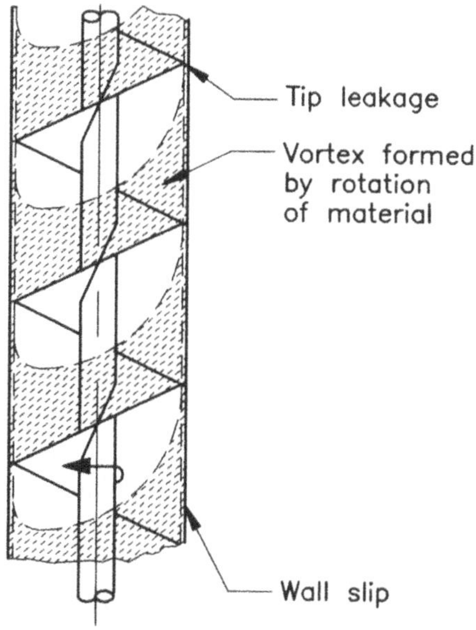

Fig. 2.2 Dynamic mode of conveying

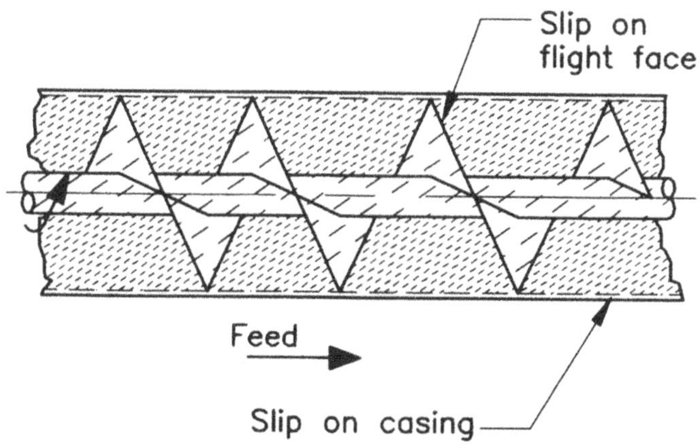

Fig. 2.3 Flood mode of conveying

sectional loading becomes highly sensitive when this level of fill is exceeded. US publication CEMA 500 provides substantial information on the constructional and performance features of standard screw conveyors. BS 4409 lists UK standard sizes and ISO 7119 IDT, ISO 1050 EQV, and ISO 1819 EQV are international equivalents for calculating drive powers.

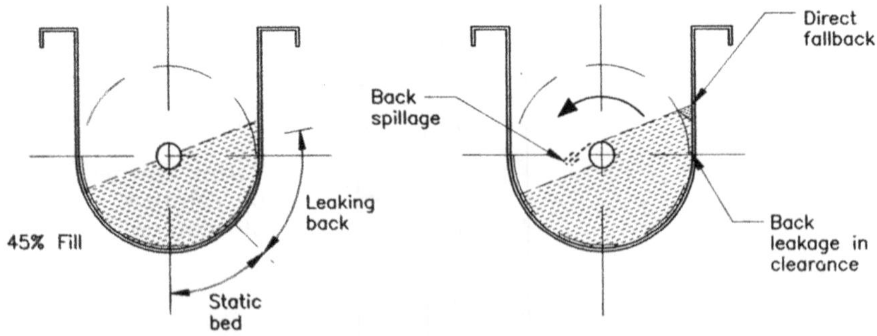

Fig. 2.4 Loading of screw cross-section in gravity mode of operation

Horizontal screw conveyors are widely used in industry for handling loose solids and pastes in a range of standard sizes. A principal feature of their operation is that the amount of material transported is controlled by prior equipment. The machines are generally constructed with screw flights of uniform diameter and pitch, and in lengths from under 1 m, to over 50 m long. Multiple inlets and outlets can be accommodated. Intermediate bearings are required for long machines, except for special applications where the screw runs on the trough or on liner plates. Where the feed covers the screw in the inlet, the machines function as feeders, and special conditions apply to considerations of power, intermediate bearings, and the influence of apparently minor variations and features of construction. Design variants can accommodate many of these effects, but the units then cease to be simple conveyors.

Inclination of the screw axis to give an elevating function has three main effects that combine to reduce conveying capacity. Firstly, the inclination of the screw face to the horizontal, on which the material has to slip to move forward, is lowered. The consequences of reducing the slope of the face angle of the screw face are more pronounced on the inner regions of the flight face, because the helix angle of the screw flight is much coarser at smaller radii than at the flight rim. The reason for this is that the flight pitch is constant for all radial positions, so the face helix has to progress the same axial distance within a smaller circumferential length. The effect of this helix variation is that inclination of the surface of the screw blade near the centre shaft rapidly causes this region of flight surface to fall below the angle of sliding friction of the bulk material. As a result, material in this region tends to rest, rather than slip, on the blade, to be carried over the shafts and fall back into the preceding flight pitch spacing as the screw rotates.

A second factor reducing the amount of material conveyed as the axis is inclined, is that the dynamic repose of the material being moved remains unaltered in relation to the horizontal. The volume of axial cross-section of material resting between the flight pitches is reduced as the inclination of the flight face approaches the repose angle, so less material can be conveyed. The effects are progressive (Fig. 2.5).

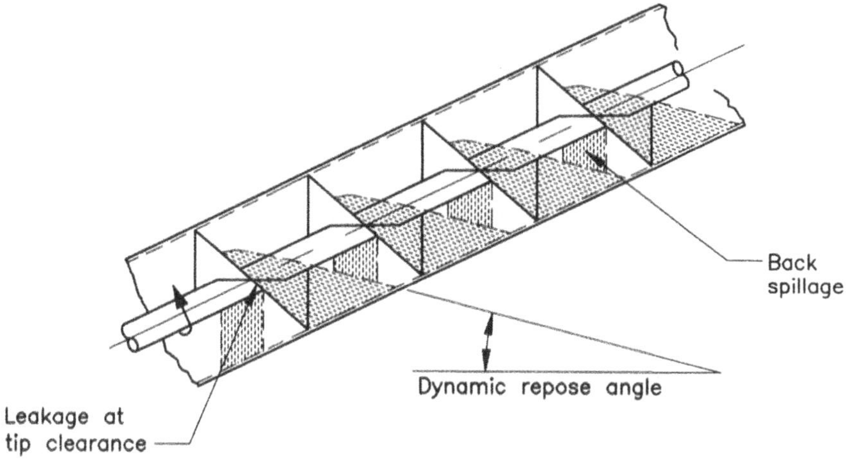

Fig. 2.5 Effect of steep inclination on gravity mode conveying

The third factor detracting from the elevating capabilities of a screw in gravity mode operation is flight tip 'leakage', i.e. how easily the material flows through working clearance gaps between the flight and the casing. Even when a screw is conveying horizontally, there is a degree of 'leakage' of material through this tip clearance space, because the different levels of material before and after the flight face allow spillage through the gap, Fig. 2.6. This effect increases rapidly at high cross-sectional loading, particularly when the level of material on the side of the casing exceeds the centre height of the screw, because the curvature of the circular screw rim of the casing progressively exposes an unrestricted overflow opening against the vertical face, see Fig. 2.7. As the screw casing is inclined, the material against the screw face tends to rest at higher levels at the side of the trough, as the contents held in the flight pitch space are biased back on to the screw blade. Increasing gravity also acts to promote 'back flow' of product through the flight tip clearance as the casing is inclined more steeply.

A further factor that amplifies back leakage when the screw axis is inclined, is wall slip of parts of the bed of material occupying the tip clearance layer. The

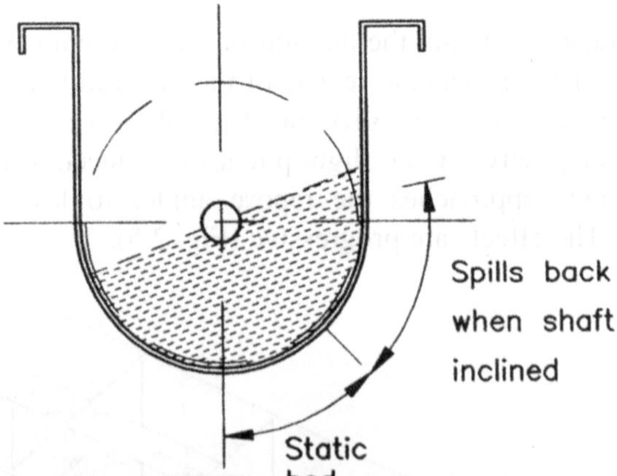

Fig. 2.6 Leakage through tip clearance

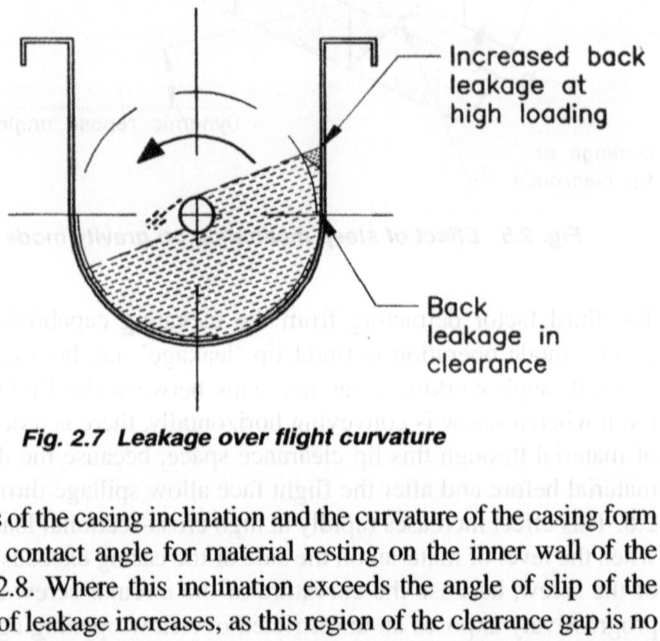

Fig. 2.7 Leakage over flight curvature

compound angles of the casing inclination and the curvature of the casing form the local surface contact angle for material resting on the inner wall of the casing, see Fig. 2.8. Where this inclination exceeds the angle of slip of the product, the rate of leakage increases, as this region of the clearance gap is no longer filled with a dead layer of material on the casing wall to restrict back spillage. When the slope of the casing exceeds the angle of wall friction of the product, all the boundary layer is able to slip backwards through the flight tip clearance, unless, as is unlikely at these angles, it is restricted by material filling the prior flight pitch space.

Short-pitch screws allow conveyors to work more efficiently at moderate inclinations, but inclinations above 35 degrees are unusual for gravity

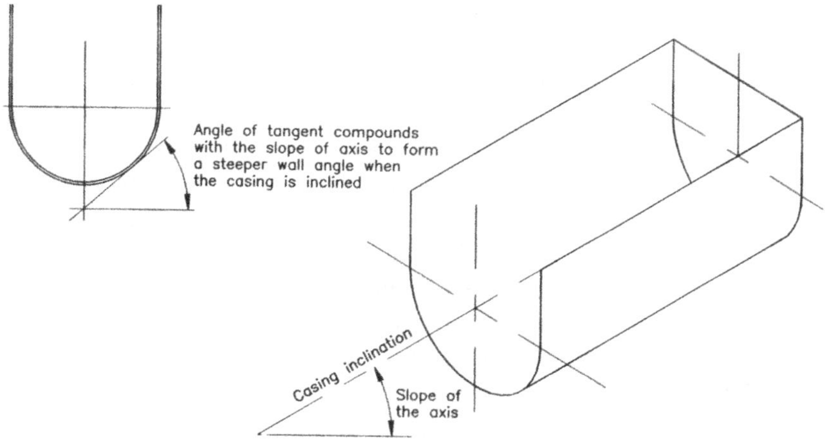

Fig. 2.8 Slippage on casing wall due to compounding of inclination with curvature

mode conveyors because of the above back leakage problem. The precise behaviour of the bulk material depends upon its flow condition and frictional characteristics in relation to the flight face. As a general guide, the 'gravity mode' handling capacity of a screw conveyor inclined at 30 degrees to the horizontal reduces to about 30 percent of its horizontal capacity, and decreases rapidly at steeper inclinations.

Steep inclinations are sometimes employed for special duties, such as de-watering, where the screw collects saturated solids from a settling tank and the liquor drains back while the solids are elevated. Back leakage, in these cases, is encouraged by the use of ribbon-type flights. Non-free-flowing materials, such as wet filter cakes, do not suffer as badly from the problems of 'back leakage', hence inclined screws can handle higher rates of these types of materials at a given angle than free-flowing materials. Loose surface moisture on some such materials also causes them to have very low wall friction values, allowing elevation at unusually steep angles. Such applications are sensitive to the properties of the material and strict verification of performance should be supported by approved trials. Short-pitch conveyors are occasionally specified to deal with difficult flow bulk materials and reversing applications. The key point is that the precise handling behaviour of a screw depends upon the material's flow properties, and on its frictional characteristics upon the screw flight.

The culminating effect of trying to elevate bulk material by means of low-speed screw conveyors at steep angles is that the fallback of product fills the cross-section of the casing at the inlet to the conveyor. Providing that further material is fed to the machine, and that the inlet hopper is of good flow form so that it will generate an overpressure to resist back leakage down the casing, then the screw will continue to extract material from the inlet and convey it up the axis of the conveyor in a flooded mode. In this form of material transfer the screws act as flood feeders, but there are various drawbacks compared with horizontal feeders. The output is relatively small and unpredictable. The infeed is only effective with free-flowing materials, and the energy input required absorbs a great deal of wasted powder entered to the bulk material by way of confined shear, leading to particle attrition and excess work input. The most unsatisfactory feature of this form of conveying, however, is that when the feed to the inlet stops, the material falls back down the screw conveyor casing and conveying effectively stops, with the screw continuing to turn while full of material.

Some of these limitations are overcome by utilizing a short screw feeder to deliver material into the inlet of a steep screw conveyor, see Fig. 2.9. Preferably this should be of the end delivery type in order to provide a direct infeed pressure. This feeder will give more accurate regulation of the feed rate, provide a reliable feed for poor flow materials, and restrict gross overloading of the inclined screw cross-section on restart from a full screw condition. However, once this approach is adopted, the inclined screw may as well be run at a faster speed to clear the incoming amount and the whole character of the application then changes.

As steep conveyors will not self-clear of product, slow-speed screws are rarely used at steep inclinations. Flood feed conditions give rise to high torque absorption compared with screws of lower cross-sectional loading. The starting torque can be particularly high, as the settled material attains much higher shear strength due to the total confinement of the casing than when in a dilated dynamic condition as conveyed. It can also take some time from start-up for the high torque loading to diminish, particularly if further material is entering the feed port to replace material carried away. Therefore, for steep screw conveyors, it is good practice to incorporate one or more increases in the pitch of the screw along the axis, after the inlet.

Short-pitch flights offer a better mechanical advantage for moving the material settled in the lower end of the casing, and the progressive pitch construction ensures that material is dilated in subsequent travel by the following flight region. This technique not only overcomes problems of

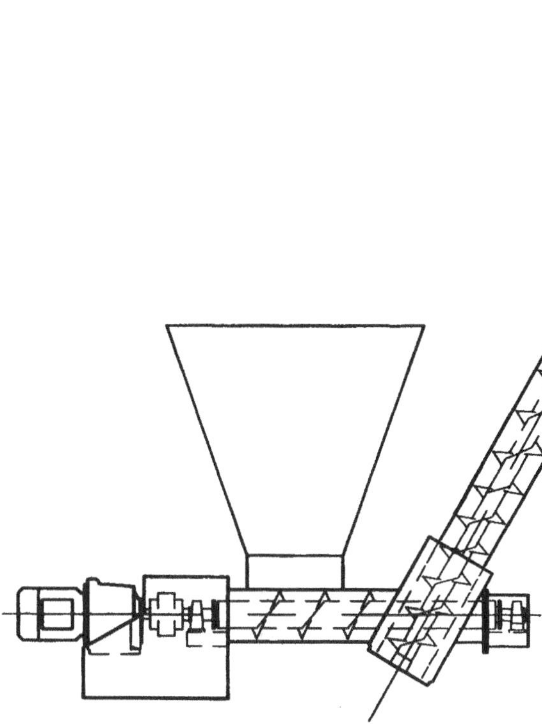

Fig. 2.9 Use of short screw feeder to steeply inclined conveyor

high torque loading when starting in full conditions, but also reduces the power needed during normal operation. Where practical, steeply inclined screw conveyors should run for a short period before fresh feed is offered to the machine. This will clear the inlet region of settled product, and allow material in the conveyor to achieve a dynamic condition before any extra load is placed upon the drive by the entry of fresh material.

2.2 Screw elevators

To overcome the inclination limitations of gravity mode conveying, screws are enclosed in circular casings and run at higher speeds of rotation. Rotation of the mass of material within the casing gives rise to outward radial pressure due to centripetal forces. Above a critical speed, at which the restraint to material rotation is developed by friction on the inner wall of the casing, the material elevates as a dynamic vortex form on the face of the screw flight. Gravity forces become negligible relative to the dynamic forces incurred, as the material is advanced in a helical path determined by the frictional relationship between the material and the face of the screw blades. The casing friction does not influence the direction of motion of the material being conveyed, and hence has no influence upon its conveying efficiency, but it does determine power needs. The casing friction has to exceed a specific value relative to the friction of the material on the face of the screw flights; this must be sufficient to prevent the mass of solids rotating with the screw, otherwise material rotates with the screw and will not elevate.

Screw elevators operate at all inclinations, up to the vertical. Intermediate bearings offer a flow obstruction, necessitate a gap in the flighting, and require access for service, so their use in screw elevators must be avoided where possible. Screw elevators' lengths are, therefore, normally limited by the stiffness needed for the centre tube, to prevent casing contact by deflection or shaft whirling. For this reason screw elevators in plant use rarely exceed 8 m overall casing length.

Apart from the danger of shaft whirling, when the rotational speed approaches critical resonance conditions, screw elevators can suffer the phenomenon of 'epicyclical rolling'. This occurs when the contact tip of the screw ceases to slip on the casing, but rolls around the inner circumference, much like a penny in a bowl. The speed at which the flight axis progresses to maintain the casing contact is determined by the ratio of the radial clearance dimension to the screw flight diameter. This is a large magnification factor to rotational speeds that often exceed 100 r/min. Excitation frequencies in these conditions are extremely high, giving rise to large acceleration forces acting on the out-of-centre screw.

The combination of large forces and high frequencies causes severe vibration. Transmission of these forces to the supporting structures often causes alarm for the integrity of the framework and even, in some cases, of the building in which the elevator is housed. The basic nature of this problem is very similar to the 'vibratory chatter' that tends to occur with long conveyor screws handling damp products, forming a strong bed of material on the casing clearance. In conveyors, the screw 'rides up' the casing wall and oscillates, sometimes alarmingly. There is, however, no comparison to the severity of vibration that can occur with screw elevators under the above circumstances – a comparatively rare, but unforgettable, experience.

In order for the material to enter the swept volume of the screw at the inlet port, it has to provide a positive pressure at the end of the inlet flow stream. The supply hopper therefore has to be of good flow design to deliver reliably, often down an inclined connecting chute, and provide sufficient residual overpressure to overcome the tendency for material to be thrown off the screw by centripetal force, see Fig. 2.10. Terminal clearance of free-flowing material at the inlet can be minimized by reducing the exposure length of the screw, either by an adjustable slide or short inlet. Output capacity is lowered by the use of short inlets; conventionally, about two screw pitches are exposed to the incoming product.

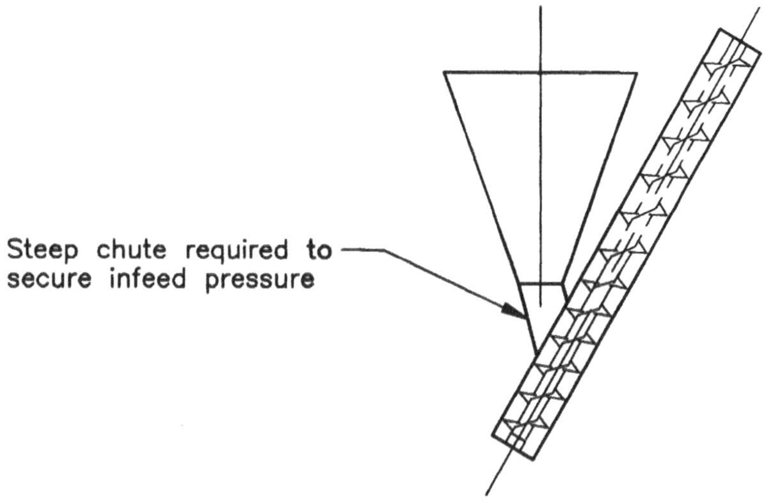

Fig. 2.10 Inlet conditions to steep, high-speed, screw elevator

It is not easy to achieve the reliable feed of a bulk material with poor flow characteristics into a steep screw elevator, by virtue of the difficulties of developing an adequate overpressure at the end of a supply flow channel. Except for very crude applications, screw elevators should not be expected to perform metering duties. Not only is the feed rate sensitive to minor features of geometry and material properties, but variation of the screw rotational speed does not provide a linear response. In fact, if the screw speed falls below its critical transfer value, the feed stops altogether, and at very high speeds the output declines, due to infeed resistance. The use of short screw feeders, directly injecting material into the inlet of a screw elevator, is therefore a useful means of overcoming infeed difficulties. A separate feeder screw offers accurate feed control, the potential for a low inlet height, and extended length to give a high-capacity holding hopper within limited headroom, see Fig. 2.11.

High-speed screw elevators self clear, up to a point. On the main run of casing the dynamic action moves forward all material that bounces

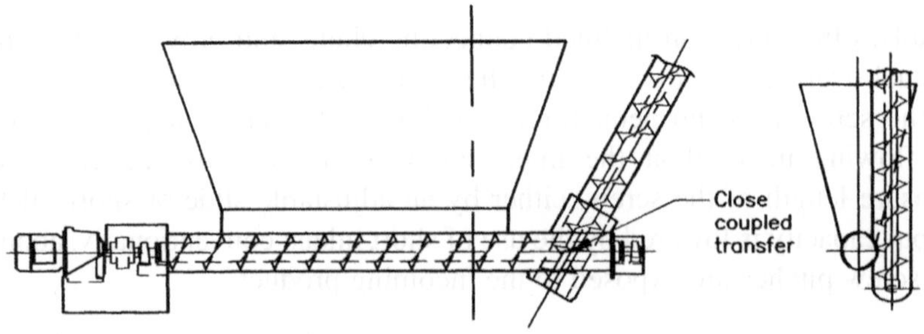

Fig. 2.11 Extended screw feeder to screw elevator

between the flight face and the casing wall. There is an essential operating clearance between the screw and the casing that allows back leakage. The amount that falls back at the end of a run is very dependent upon the nature of the material. Fine powders tend to be agitated to a fluid condition as the amount in the casing reduces to allow substantial dilatation. This fluid product falls back into the inlet region as the screw slows to a stop. At the inlet, once the feed ceases, the material is flung from the screw blade back into the inlet chute. Some re-enters the screw form, to be thrown out again. Ultimately a stable amount of dynamic residue remains, to settle to a firmer bed around the screw when rotation ceases.

It is not exceptional for a long, steep elevator conveying a fine powder to experience a cyclic oscillation of behaviour at the end of a batch run. Whereas the machine may operate smoothly, continuously transporting away all material entering the hopper until the infeed ceases, once the hopper empties and the inlet region clears, the material in transit is no longer 'backed-up' by following product.

The sequence of events which then transpires is initiated by a progressive dilatation of the product through the length of the casing, with discharge from the outlet progressively declining as the effect works up to the elevator outlet. As the residual contents in the casing become aerated they achieve a more fluid condition. Eventually, the degree of looseness is marked by a rapid flush of the contents from the casing back into the feed hopper. The 'fallback' material initially partially fills the supply hopper in a fluid state. As the material settles and attains a less fluid condition, it is re-entrained by the screw to move again up the casing. Before reaching the outlet, the inlet section exhausts the supply of product and the material in the casing is progressively re-dilated as excess voidage formed in the bulk works its way up the casing, until once more the material becomes totally fluidized and runs back down into the hopper.

This cycle, of in-feed and fallback, will continue until the elevator is stopped, at which stage the material will run back into the hopper to settle. There is a danger with large or long elevators having small feed hoppers, that the amount of material in transit, and hence the fallback volume, can be greater than the hopper will hold. Material then disconcertingly spills over the rim of the hopper as it first flushes back from the casing and overfills the hopper. Some materials which fluidize in this manner settle to a firm, poor flow condition. Whereas they may feed well when initially poured into the supply hopper, the same material may present serious flow difficulties when finally settled at the end of a cycle, as already described. A further operating hazard for screw elevators handling material that settles to a firm condition, is that of stopping the elevator with infeed product in the hopper. This situation prevents material running back out of the casing. Material settled in the lower portion of the casing will then offer a high resistance to re-starting the elevator, unless design provision is made by way of incremental flight pitch construction of the screw auger.

Screw elevator design entails considerably more expertise than that required for screw conveyors. The performance is more sensitive to the nature of the product handled, and a host of operating hazards, not relevant to screw conveyors, may be experienced. Screw feeders incur different, but similarly specialized, experience for optimum design exploitation.

2.3 Screw feeders

2.3.1 Common factors

The term ‘screw feeder’ can be applied to any screw-type machine that controls the rate of feed that it dispenses. This control is given when the inlet section of the screw is covered with bulk material. The geometry of the screw and its speed of rotation then determine the volume of material extracted from the supply channel. This situation may be temporary, as the ‘feeder’ serves to limit the amount of material discharged only during the times that the inlet is full of product. Machines of this type act as temporary screw feeders, while the in-feed conditions persist. Nevertheless, the fact that they then control the output rate at some times has implications relevant to their use and to other equipment.

2.3.2 Differentiating factors

Dedicated screw feeders, however, are designed to operate with a permanently full inlet region. The many differing forms of screw equipment that fall into these general classifications of both ‘acting’ and 'dedicated' screw feeders can be differentiated according to their primary design function, as set out in more detail in Chapter 3. The key performance elements of any screw feeder are:

(i) the form of the supply hopper, particularly the interface section with the screw; and
(ii) the geometry of the screw section exposed at the interface to the hopper.

These features interact to determine the drawdown characteristics of the solids feed channel into the screw. The flow regime generated not only has an important influence on the reliability of the feed and the consistency of condition of the bulk material, but also affects the form in which the contents discharge from the hopper serving the feeder.

The amount of material extracted from a flooded region by a screw feeder is mainly determined by the screw construction at the outlet end of the infeed region interface. The restraint offered by material in the clearance space between the flight tip and the casing as the screw leaves this region affects whether extra material is carried out by the boundary layer. This section of casing adjacent to the 'exit' point of the screw from the supply hopper is termed the 'choke section'.

In most cases the choke section is of circular form and extends for a length of at least one screw diameter along the casing. Its purpose is to prevent material being expressed from the hopper over the screw form at an indeterminate rate. If the casing is of 'U' cross-sectional shape, a 'saddle'-shaped portion is often fitted under the flat cover to fill the space above the screw. A flat 'choke plate' across the casing adjacent to the hopper end wall is an economical alternative, but is less effective in preventing material in a dilated condition from 'flushing'. A longer choke section than normal is appropriate where the bulk material is in, or acquires in transit, a 'loose' condition.

It should be noted that no screw form or length of choke section would prevent the escape of a fully fluidized material. Apart from the essential tip clearance space between the flight and the casing, there is also an unrestricted channel of flow around the helix form of the screw flight. This channel will allow the passage of a fluid-like bulk material, which is invariably under the pressure of a hydrostatic head if the material at the base of the stored contents is in a fluid state. It is quite impressive to witness a fine powder product squirting from any pinhole, crevice, or non-watertight seal when the bulk material attains a fluidized condition in a container. The author has observed unrestrained flow of hydrated lime through an inclined screw feeder over 5 m long, followed by a horizontal screw conveyor 10 m long, continuing after both the machines had stopped. This event followed discharge from a silo being aided by the use of fluidizing pads.

The distance from the hopper interface to the outlet of the casing must be long enough to prevent the material from trickling through, when the feeder screw has stopped. The 'angle of repose' of a material dilated by moving at a high rate through the flow channel may be very low, allowing the geometry of a screw to offer a 'line of sight' down one side of the axis, for a full pitch length or more. Taking account of the clearance between the screw and the casing, it is good practice to allow a length of casing equal to at least one-and-a-half screw diameters between the end of the hopper and the start of the outlet, to avoid through-flow.

Screw feeders usually work in a horizontal position. The main reason for inclining a feeder is to overcome headroom problems but it is not recommended to exceed 15 degrees inclination with a metering screw. Angles up to 30 degrees are sometimes used for special duties, as previously described in the conveying section. The design of screw feeder hoppers, and screw inlet geometry for inclined duty, requires special consideration – as described later. The two main forms of dedicated screw feeder may be classified as:

(i) those used for discharging from a bulk storage container; and,
(ii) those whose prime function is for dispensing material at a controlled rate.

There is no clear demarcation between these applications, as bin dischargers may be of small scale and have volumetric or gravimetric feed control, and some metering feeders are equipped with bins of large storage capacity. Design variations that include extended delivery lengths, and/or inclined, may be described as 'bin feeder–conveyors', or 'elevating-feeders', to further blur these classification boundaries. However, for convenience, dedicated screw feeders may be grouped according to their prime function.

2.3.3 Bin dischargers

Screw feeders used to discharge from bulk storage are normally designed as an integral feature of the facility, see Fig. 2.12. Their essential function is to provide a reliable discharge of the stored contents. The favourable flow form offered by wedge-shaped hoppers, facilitated by a slot outlet, is often a main reason for using a screw feeder. A second common objective

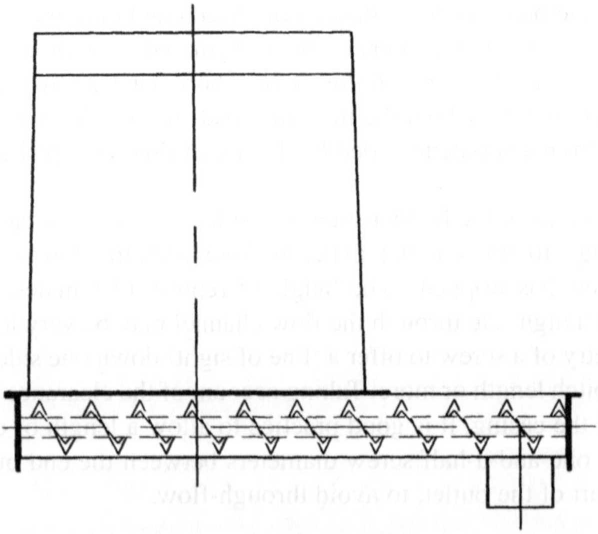

Fig. 2.12 Bin discharge feeder

of fitting bin discharge screws is the enhancement of storage capacity, by means of extending the outlet of a storage hopper to a long slot.

In cases where a single screw width will not provide an adequate width of opening to avoid arching, or extra storage capacity is needed within a limited height, a broader outlet width is obtained by the use of multiple screws in parallel, see Fig. 2.13. For either reason, design emphasis is on maximizing the hopper opening size, and thereafter directed to providing efficient extraction over a long, exposed screw length. To secure the flow benefit of a slot, live flow should take place over a length at least three times the width of the slot.

Longer slots, however, usefully convert the shape of the lower section of a storage vessel from converging end walls to vertical end walls. Flow characteristics are thereby improved, by changing from convergence in two planes to the single plane convergence offered by a 'vee'-shaped hopper. The principle advantages of this feature are, that the elimination of gullies and exploitation of uniaxial convergence allows the wall inclination to be reduced, either for mass flow design, or for the ultimate self-clearing wall inclination of a non-mass flow hopper. For use with free-flowing materials that do not require a mass flow pattern of flow in the storage vessel, a long, 'collectable', screw form may be adequate. Selective increments of extraction are usefully employed to reduce power needs for the feeder drive, and reduce the size of static regions of storage in the hopper.

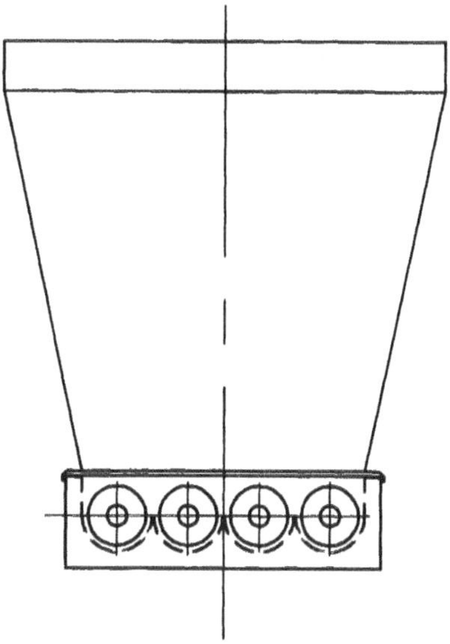

Fig. 2.13 Multiple screw feeder

The facility to transfer the discharge to a position offset from the centreline of the storage vessel is often convenient, as is the ability to control the outfeed rate and provide remote starting and stopping. Thus, multiple advantages may be gained by the use of discharge screws. Units can be inclined at relatively shallow angles, to secure some recovery of headroom or provide a limited elevating capability. The section of casing beyond the choke section may also be made longer, to provide an extended conveying run (Fig. 2.14). The length of span that the centre shaft or tube of the screw can accommodate without undue deflection, limits the overall casing length of a feeder. Intermediate bearings are to be avoided on screw feeders, because material filling the whole cross-section has to be pushed across a gap in the flights, past the obstruction of the support bearing. Blockages, and a range of other difficulties, can then arise.

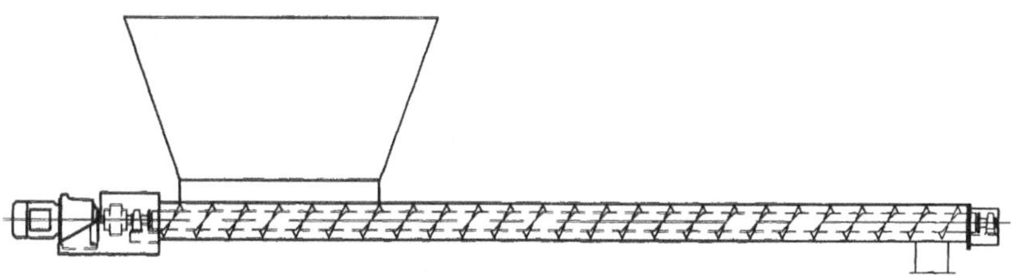

Fig. 2.14 Screw feeder with extended conveying section

An important feature of the design of a bin discharge screw, is the pattern of flow that it generates through the interface to the storage container. For a bin of mass flow design, it is essential that the total area of the outlet is active when the screw is extracting material, otherwise there will be a dead area that prevents a zone of the hopper from moving, whatever the hopper form. To provide continuous extraction over the full axial length of the feeder, the screw geometry must be such as to generate a progressive increase in transfer capacity. It is not necessary for increments of extraction capacity per unit length to be completely even. This is rarely practical. The section of screw furthest from the outlet will move material away from this end, and refill its full cross-section from above. All other regions can only absorb the incremental transfer capacity of the screw, as material from prior regions moves along the axis. However, this initial entrainment region at the start of the screw encourages flow down the end wall of the bin.

Variations in screw geometry, and the differing extraction patterns these infer, are not strictly reflected in the flow pattern of material from the hopper. In practice, a potentially uneven flow channel will itself provide a degree of attenuation for the differing theoretical extraction rates, by virtue of the shear strength of the material. Screw filling pressures are transferred from areas of high flow velocity, to adjacent regions where the flow velocity is lower. Filling pressures are thus increased where the extraction geometry provides low increments, and decreased in regions of the screw that tend to extract more material. The local flow channel where the geometric extraction volume is high, is also more dilated than the region where the flow rate is lower. The combination of the increased density condition and higher filling pressure in regions of low extraction, is to modulate the density and filling efficiency, to redress partially the theoretical difference in extraction rates, and give a more even inflow along the screw axis than the geometry would suggest.

2.3.4 Metering screws

In contrast to the emphasis placed upon the enhancement of storage capacity with bin discharge screws, feeders designed for controlled-rate dispensing have the prime objective of securing reliable, predictable, and controllable solids flow. Two conditions have to be satisfied: first, to ensure the flow channel is smooth and continuous in all conditions of operation; second, that the volume of the feed screws fills in a consistent manner with loose solids in as uniform a state of density as possible. In general, a mass flow channel best serves both objectives, due to its performance reliability and because the total flow channel passes all the material through a similar condition of flow stresses. Variations of bulk condition are minimized, whether resulting from segregation, or by dilated material reaching the screw through preferential flow channels. The overriding objective of filling the screw to a consistent condition, dictates the characteristics of the screw, the hopper interface, and the hopper geometry. Essentially, the material must enter the swept volume in a consistent state, and the overpressure exert a uniform packing pressure on the contents. This generally means that the screw form is exposed over a length of between three and five diameters, and generates progression extracting capacity.

Supplementary equipment, such as agitators, vibrators, air cannons, wall-activated devices, and sundry methods of injecting air, are commonly used to promote flow. While these can overcome bridging problems, and in some cases act to de-aerate fluidized powders, they introduce uncertainty about the form of flow channel developed in the hopper serving the feeder, and must be proven empirically. When a mass flow pattern is a requirement of the supply hopper, the use of agitators, air injection, or fluidizing pads to promote flow should be avoided, as the pattern of flow they generate cannot easily be predicted.

Chapter 3

Screw Feeder Types

Screw feeders are normally subjected to flood feed of a bulk material over the exposed inlet section of the screw. The volume of material moved by the feeder is then dependent upon the speed of rotation of the screw. In a wider sense the term 'screw feeder' applies to screw conveyors that normally operate with their feed rate controlled by prior equipment, but act as feeders by controlling the rate of discharge when material in the supply channel completely covers the screw at the inlet point. These circumstances apply in hoppers subjected to surge loads, batch fills, or products accumulated in the supply hopper if the screw is stationary for a short period. Such applications are here termed 'collecting screw feeders', and 'screw conveyor feeders', but considerations relative to screw feeders of all types bear strongly upon the design of such equipment and their feed channels.

3.1 Collecting screw feeders

Collecting screw feeders are installed to discharge bulk material that is delivered by other devices over an extended length. Typical applications are screw units fitted into dust-collecting filter units, under dump hoppers, collecting from the width of rotary filters, the presses of plate filters, or batch drops from mixers or other process equipment, as typically illustrated in Fig. 3.1.

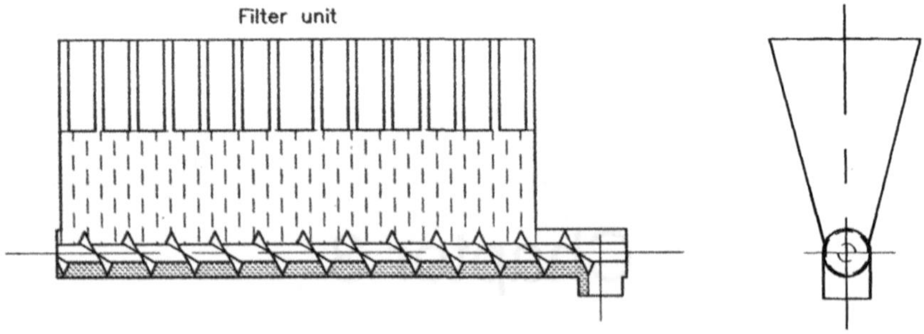

Fig. 3.1 Collecting screw feeders

In some cases the normal condition of loading on to the screw is continuous and at a rate that the screw can clear without the local cross-section of the screw being covered, but on receipt of a surge in the feed rate, material accumulates to cover the screw cross-section. The unit will then discharge the accumulated material at a rate determined by the screw capacity. Any equipment in the ensuing conveying route must be capable of dealing with the delivery rate prevailing under these conditions. The usual form of screw flight used for a collecting feeder is one of uniform pitch construction throughout the whole length of the screw. This type of screw is simplest and most economical to manufacture, but has two disadvantages. Extraction only takes place along a short length of a flood-fed screw and, because this leaves stationary material over the screw, in other regions, more power is taken to shear through the bed of static product where there is no extraction.

A uniform pitch flight construction empties the contents from one end of the collecting hopper, leaving the remaining level of material undisturbed until the extraction has worked its way along the screw, Fig. 3.2. Any subsequent batch of fresh material dropped into the hopper invariably finds its way into the drawdown channel formed by the material, to discharge before the original load. The residual portion of the original stock may deteriorate in various ways, thereby causing production or quality problems.

High starting loads are especially expensive if variable speed drives are used, because of their limited starting torque characteristics, compared with direct-on-start squirrel cage motors that develop up to two to three times normal full-load torque for short periods on start-up. Similar considerations apply to hydraulic and pneumatic drives, the strength of safety clutches and shear pins fitted to safeguard the equipment.

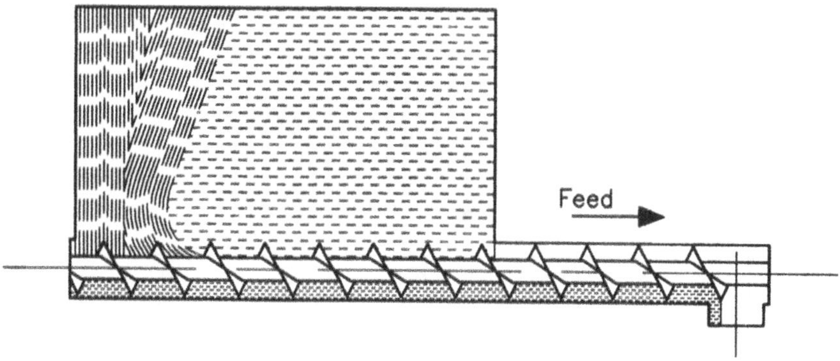

Fig. 3.2 Preferential extraction of uniform pitch screw

Furthermore, significant power in excess of normal running requirements causes other problems. In the case of an overload, such as the outlet of the machine becoming blocked, surplus power installed to overcome a starting load dispersed over the length of the feeder will be concentrated upon the final screw flight that is delivering material up to the blocked outlet.

It must also be recognized that the specification of generous power allowances to overcome exceptional loading conditions, or give a margin in cases of uncertainty, will provide a surfeit of power, to add to the temporary overload capability of the drive unit, and act on any local resistance encountered. In general, the mechanical strength of a full-bladed helical screw construction is relatively high in relation to the duty falling on a local region. The torsion and bending strengths of ribbon-type screws and centreless screws, as described in Section 4.3, are considerably less than the full-bladed variety; therefore, more care is required when selecting a drive for such types of construction.

There are various ways to reduce the power requirements of a screw passing through a bed of bulk material. The use of a short pitch construction throughout gives an improved mechanical advantage for moving the material. The torque is correspondingly less and the reversing capacity of the screw is not impaired, should this be required. The rate of transfer is reduced, but selecting an appropriate speed of rotation can compensate for the lower rate. However, a screw of this form will still only extract from the first one or two exposed pitch lengths. In order to spread the extraction length along the axis, one or more increases in pitch along the screw axis provide increased transfer capacity and, therefore space for more material to enter, Fig. 3.3. This construction also reduces the torque taken by the screw, as it is easier to shear the flowing material at the region of inflow, than to shear static bulk material over an unchanging screw form.

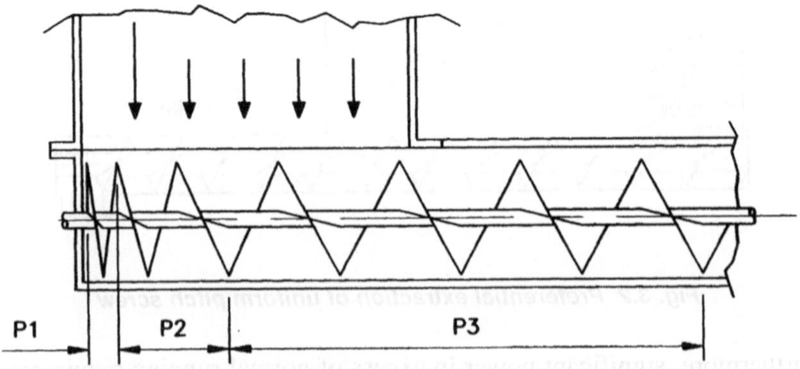

Fig. 3.3 Stepped pitch construction

The ultimate design is one where the screw pitch offers a continuous increase in capacity along its length. The range of pitch variations that are practical limits the extent to which this desirable feature can be implemented. Very short pitches result in the flights being close together, forming a deep pocket between the screw blades. As a result of frictional drag within narrow blades, the transfer efficiency of the screw is reduced and material in the bottom of deep channels tends to rotate with the screw. In extreme cases, the full cross-section becomes clogged – a condition called 'logging'. An extended pitch construction is less efficient at moving the contents. Overlong pitches transfer less material, because the material will rotate more with the screw and absorb more power.

The optimum pitch length for maximum screw handling capacity, with a particular screw geometry of blade and centre tube diameter, is dependent upon the sliding friction between the face of the screw flight and the product being handled. As the pitch of a screw is increased, the efficient transfer capacity generally reduces steadily from the 'swept volume' calculation of screw area multiplied by pitch. Transfer efficiency deteriorates rapidly when the ratio between the pitch length and the screw diameter exceeds unity. The direction of material movement progressively increases rotation around the screw axis, until the stage is reached where the material rotates completely with the screw. Calculations of incremental extraction should therefore never be made on the basis of pitch change alone, particularly when the equipment is handling material with a high contact friction.

The area of a screw blade that promotes the advancement of material resting between long-pitch flights is relatively small in relation to the length of the column of material between the blades, Fig. 3.4. As a consequence, deformation forces acting on the annulus of material being moved along the screw axis can give rise to high shear forces in the section exposed to stored material in the hopper. Strong frictional resistance can also be developed on the boundary casing under the screw, in the choke section, and on the casing walls of subsequent transfer lengths of the feeder.

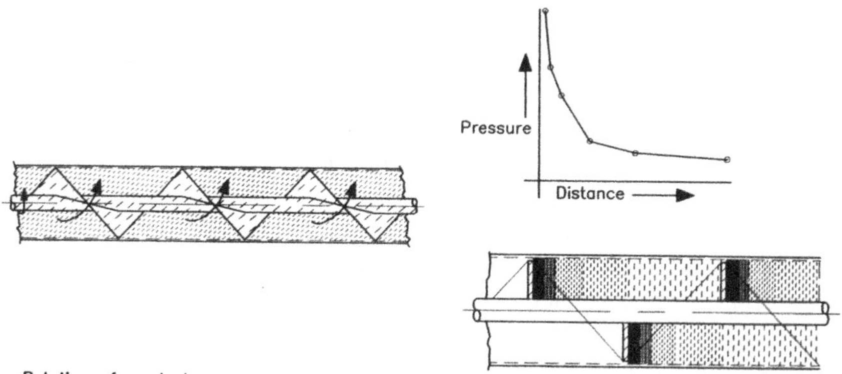

Fig. 3.4 Effects of overlong flight pitch construction

In general, pitch variations outside the ratio of one-third up to equality with the screw diameter should be avoided except for special designs. Among the consequences of overlong pitch construction are that material is not extracted from some regions of the supply hopper and, more seriously, that the material in transit may be pushed into a section incapable of transferring the full amount advanced. The product is then either compacted, or material is expressed from the screw back into the storage container. In either case, this results in high power consumption and poor feeder performance.

Whether the screw pitch construction is uniform, increases in steps, or is continuously changing along the screw length, it is essential that the manufactured item does not have any section where the pitch is less than any prior section along the axis. Poor tolerances in fabrication are a common source of operating problems. Material in transit along a screw filled within its cross-section, which then passes into a screw section of the same diameter but with a shorter pitch, will be subjected to compaction in confined circumstances. The result is almost invariably excessive power consumption and/or compaction of the material to a poor flow condition. The effect of apparently minor features of fabrication and surface finish can also have similar consequences. Projecting butt joints on segmental flights, protruding welds, or weld splatter, will interfere with smooth surface slip and can lead to blockages and/or high power losses due to the reduced transfer capacity of the local region of the screw causing compaction of the material. Uniformity of pitch construction and these constructional features are of particular importance for screws that are required to deliver to either end of the casing by reversing their direction of rotation.

The starting torque taken by submerged screws is largely dependent upon the shear strength of the material. This strength is a function of the material's nature and the loading imposed on the shearing surface. Whereas the strength of many fine materials does not alter significantly between initial failure and sustained shear, the commencement of shear in a hard, granular, confined product has very significant differences, as described in Section 1.2. To avoid extreme starting loads, discharge screws that deal with hard-grained products require that close design attention be directed to the interface region.

One way to avoid high starting loads due to initial loading conditions or time consolidation, is to start the machine before the material is first dropped on to the screw. The prior formation of a shearing layer will tend to ease re-starting, compared to a fill state with the screw stationary. A heel of material left in the hopper will prevent the formation of fresh loads directly on to the bare screw. The development of a flow condition in the hopper contents during the initial loading of the bulk material, is also a beneficial technique for aiding flow. The withdrawal of only a small volume of material, before high compacting loads are created, is enough to establish this flow condition. A small depth of product over the outlet then shields the region from impact forces and deposition conditions, which would condition the bulk to a worse flow state than one prevailing after flow has been initiated. Another way to reduce the drive power is to shield regions of the screw with inserts. These inserts proportionally reduce the area of material to be sheared and reduce the overpressure acting on the remaining exposed sections of the screws by supporting part of the hopper contents on the cross walls or offering extra wall frictional resistance. A further benefit is to reduce the degree of local confinement of the bulk material and provide a void region into which local shear expansion of the bulk can obtain relief.

A cross insert of this type gives considerable support to the flat side faces of the hopper as a structural member. It is, however, necessary to ensure that any gully angles formed by such inserts allow the contents to slide clear. A favourable way to overcome this difficulty is to use a form of cross insert with vertical walls at the lower end, Fig. 3.5. The gully angles form part of the collecting section of the hopper. The section employed to contain the settled material comprises a number of 'vee'-shaped hoppers with vertical end walls. Longitudinal inserts, as shown in Fig. 3.6, cover the full length of the screw and provide a degree of overpressure relief. They are also used to modify the flow regime of hoppers by providing twin flow channels of differing geometry to the original 'vee'-shape.

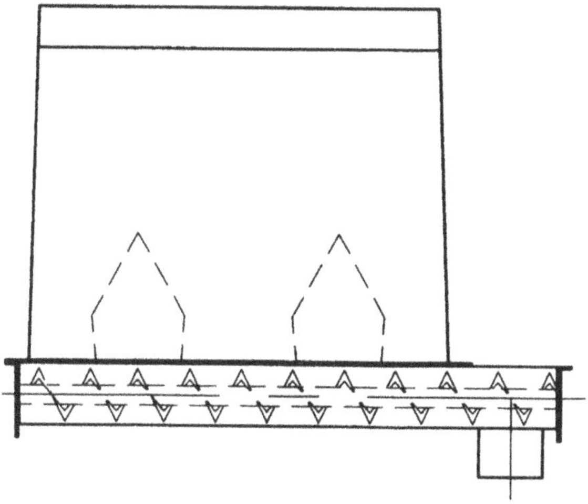

Fig. 3.5 Cross inserts

Allowance must be made for the differing extraction capabilities of the two sides of the rotating screw. Contrary to intuition, the rising side of a rotating screw allows more material to enter than the descending side. This is because the face of the rising blade is uncovering a partial void as the working face moves material away. For this reason, twin screw feeders should work with contra-rotating screws that move in a downward direction in the central region, rather than at their outer sides. This arrangement encourages maximum drawdown from material against the side walls of the feed hopper. Spreading the flow channel in this way better matches the cross-sectional area of storage, encourages slip on the walls, and compensates to some extent for the tendency for a preferential flow channel to develop in the centre of a 'vee' hopper. This tendency results from the preferential fall of material from vertically above the outlet, and the fact that the area from which the outlet draws expands sideways where material is partially resting on the container walls.

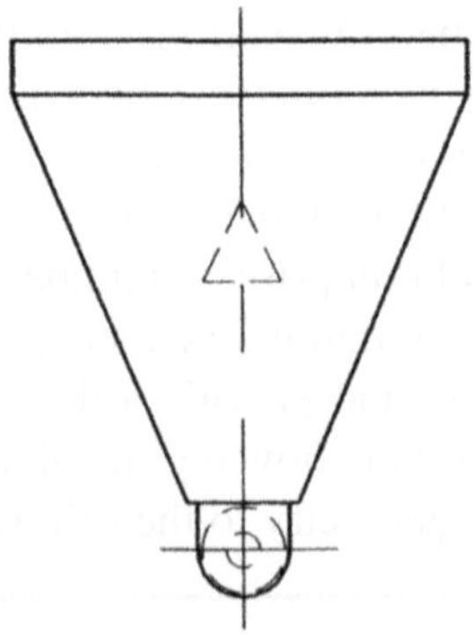

Fig. 3.6 Longitudinal inserts

If twin screws are fitted to enhance storage volume, rather than providing a wide outlet for flow, then the capacity can be further enhanced by the use of a central insert ridge between the two, to give a wider base to the hopper section, Fig. 3.7. Although the feeder outlet port is made wider by this technique, the discharging material can be focused by means of an inclined chute at a much lower inclination than that required to stimulate wall slip in the confined flow circumstances within the hopper. This is because flow in the discharge chute is unconfined.

In respect of the hopper form, the design requirements for wall inclination and opening width for a collecting hopper differ significantly from that of a flood-feed hopper. In the case of the former, duty material slips on the walls without confinement; therefore, the walls only need to be sufficiently steep to overcome the angle of wall friction. The minimum outlet size required for unconfined flow is, in the limit, that width of opening large enough to pass the largest lump.

By contrast, a flooded hopper has to be capable of initiating flow through the opening, and the opening size has to be adequate to pass lumps without danger of them combining to form a mechanical blockage. Although this condition is a relatively trivial problem to avoid, the determination of a critical opening size necessary for the confined flow of irregular-shaped lumps from basic measured data, is beyond the current state of the art. The minimum orifice size for reliable flow depends upon the size, particle size distribution, shape, contact surface conditions, and density of the constituent particles, as well as their manner of packing in terms of both individual and bulk isotropy. The size needed is also influenced by the orifice shape, the inclination and frictional characteristics of the walls

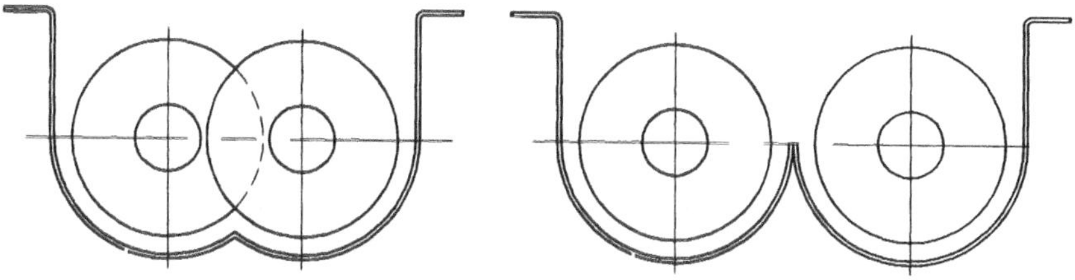

Fig. 3.7 'Close-coupled' and 'spaced' twin screw arrangements

adjacent to the opening, the velocity of flow, and the state and dynamics of the ambient fluid in the voids of the particle array.

The probability of the particles coming together to form a stable structure over the orifice is a stochastic process, but is sensitive over a comparatively narrow range of opening size difference. A flow channel slightly larger than that at which stoppages occur, may be subjected to wide flowrate fluctuations as dynamic and unstable arches erratically form and break down to impede the smooth flow of material from the outlet. With so many variables it is invariably easier to determine the required size empirically, or to make a generous allowance, than try to calculate a 'safe' value. As a general 'rule of thumb', a slot opening wider than four times the maximum lump size is usually adequate to allow stable flow through a slot if slip is taking place on the walls. An opening 25 percent wider will normally suffice if wall slip is not taking place. A circular outlet should have an opening diameter roughly double the width required of a slot, to give similar flow reliability.

The outlet of a hopper also has to overcome any tendency for the material to arch cohesively. It is likewise not practical to assess a 'critical orifice size' to avoid cohesive arching from primary particulate data. It is not practical to assess a 'safe' orifice size to guarantee the flow of a cohesive product, and no simple 'rules of thumb' exist. A comprehensive procedure, based upon shear testing, is essential to calculate an orifice size that will resist arching and 'ratholing' due to the bulk strength of the material. Figure 3.8 shows the general nature of flow of differing particle size materials through an opening of a given size, 'D'.

The calculation for the minimum size of opening to secure reliable flow and avoid both mechanical and cohesive arching is strongly influenced by whether the material slips on the approach walls. Without wall slip the outlet needs to be approximately twice as large to ensure flow. An 'expanded flow' form of hopper construction, as shown in Fig. 3.9, secures the reliable outflow benefits of mass flow, without sacrificing too much headroom. This shape of hopper utilizes a mass flow construction for the outlet region of the container. The hopper wall inclination is relaxed to a shallower angle when the span between the walls does not constitute a potential blockage problem if wall slip is not taking place, either from arching or from 'ratholes' forming in cohesive products.

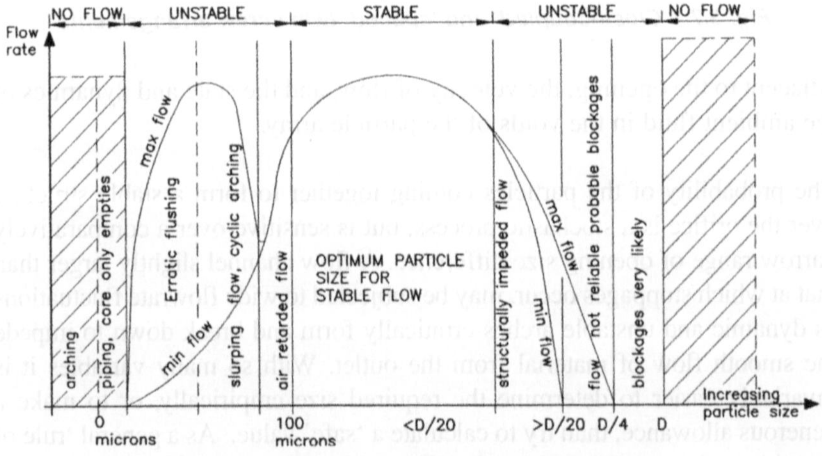

Fig. 3.8 Flow rates of differing sized particles through an orifice of a given size, 'D'

There is an alternative reason for using an 'expanded flow' type of construction. Mass flow construction requires steeper walls than those needed to provide 'self-clearance' of hopper contents. In cases where mass flow is essential to prevent contents having an indefinite period of storage but refills are of a batch nature, as with tanker deliveries, the massed flow portion of an expanded flow construction may be adequate to contain the residual working contents prior to a new delivery. Whereas the form of flow of the fresh material will not strictly follow a 'first-in–first-out' pattern, the original contents resting in the mass flow section of the hopper will not have a static region by-passed by fresh product. Indefinite storage time is thus avoided and headroom is saved by means of the composite constructional form of the hopper. To preserve the same sequence of discharge as the order of fill, it is important that the storage unit is not refilled before the remaining contents move into the mass flow section of the container. A third use of 'expanded' construction is to collect product from a wide cross-section, as from a filter, while discharging in the order of product deposition and avoiding 'dead' pockets of material. In all such cases, the transition point

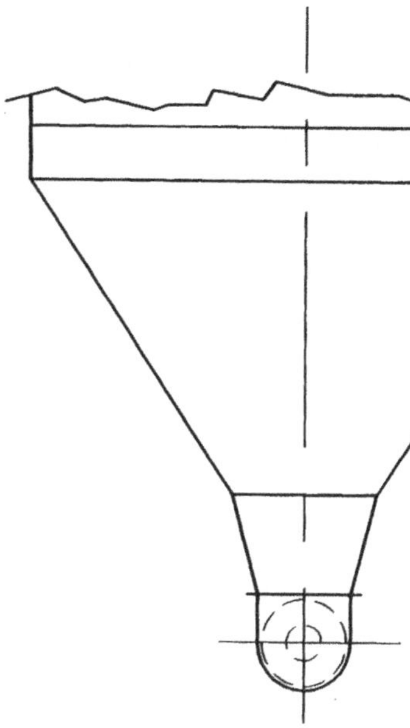

Fig. 3.9 'Expanded flow' shape of hopper construction

at which the hopper wall angle changes has to be based upon the specific criteria for the shape selection.

3.2 Screw conveyor/feeders

If the inlet of a screw conveyor is subjected to flood-feed conditions it is good practice, except in the case of a short, single span unit, to incorporate a short section of half pitch flights under the inlet and for a distance of at least one screw diameter beyond. This will constrain the loading of the subsequent conveying length to a nominal 50 percent of the cross-sectional area. A 'choke' plate or 'saddle section', Fig. 3.10, should also be included if the casing is not of circular construction, to prevent excess material being carried forward with the contents of the screw-swept volume to create a higher screw loading. As previously described, the use of intermediate bearings on conveyors with high cross-sectional loading can cause a variety of problems and are best avoided.

3.3 Bin discharge screw feeders

Dedicated screw feeders may be classified according to the primary function they undertake. These duties may be broadly broken down between bin discharge feeders and metering feeders. There is no sharp boundary between these duties as a screw installed to perform bin discharge duties also controls the rate of feed to a subsequent process or receiving point, and many metering screws are served by a significant bulk storage facility. The distinguishing feature between these classes is whether the main purpose of the equipment is to complement the storage capacity or to dispense bulk material at a controlled rate to subsequent plant. Refinements of design which develop from this primary function also tend to clarify the identification of a feeder with one or the other of these classes.

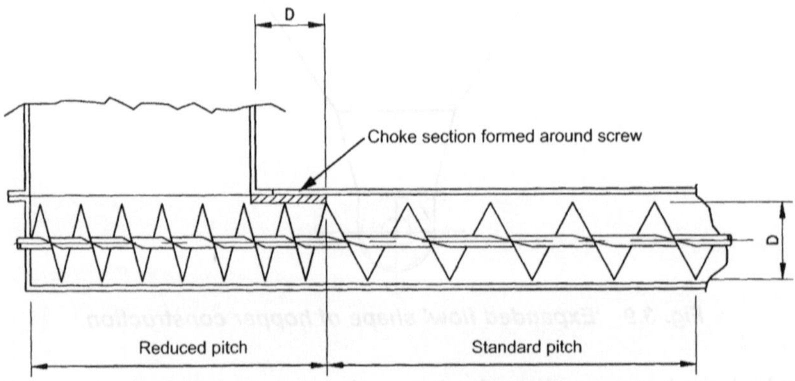

Fig. 3.10 Typical 'screw conveyor/feeder' inlet

Screw feeders for a 'bin discharge' duty are normally fitted either to enhance the capacity of a storage hopper or to aid the discharge of a difficult flow material, or both. They achieve these objectives by providing an extended length of slot outlet and a larger size of opening than are practical for use as an unrestricted bin outlet. By contrast, the storage capacity of a metering screw feeder is usually incidental to the task of providing an accurate feed to a subsequent process or delivery point. The difference is not academic, as the accuracy, uniformity, and consistency of the bulk 'state' demanded of a flow channel to serve a metering duty is usually more onerous than that called for in bin discharge duties. 'Bin discharge' feeders are normally only required to provide reliable discharge of the stored contents at a nominal rate.

For a bin discharge duty, an important feature of the discharge function is the sequence in which varied regions of the stored contents are extracted. The geometry of the screw controls the extraction pattern that is imposed upon the stored material presented at the hopper/feeder interface. A key performance factor of the feeder is, therefore, how the screw form varies along this section. A further function commonly performed by bin discharge screws is to deliver the material to a required location. In this respect, design considerations may require the equipment to have an extended length or to be reversible, in order to comply with the best site arrangement. When the length of screw is considerably extended beyond the bin outlet section it is good practice to further increase the screw transfer capacity after leaving the feed hopper, in order to avoid high power losses from moving the bulk in severely confined conditions.

The requirement for the screw to reverse severely compromises the form of screw geometry that may be employed and, by its relationship to the flow pattern from the hopper, also bears heavily upon the bin shape in its entirety. For mass flow applications the exposed length of a reversing screw feeder is restricted to about one and a half pitches of the screw, in order to avoid dead regions of extraction at the hopper interface. Longer exposed lengths of screw can serve non-mass-flow discharge applications, provided that the screw construction is perfectly uniform along its length to avoid compaction blockages. The region of hopper from which material is extracted will change from one end of the outlet to the other when the screw reverses, Fig. 3.11.

Bin discharge screws are intended to work in flooded conditions. Starting loads and extraction pattern are related, in that pressures developed at the screw interface, and the pattern of extraction imposed at this region, depend upon the design of the hopper section of the bin and how it operates. The selection of screw size, its form and geometry, and the hopper/screw interface design are interdependent within the overall bin design, for both cost and performance.

3.3.1 Left- and right-hand screw feeders

Left- and right-hand pitch screws on a common shaft are commonly used to feed material out to a central or intermediate location along the screw axis. For the screw to act as a controlling device, an insert is fitted over the casing outlet, and is sufficiently extended at each side to prevent material running through the repose slope, Fig. 3.12. It must be recognized that material is extracted from each side of the hopper according to the portion of screw that is covered. If there is a bias in the volume of contents above each side of the screw, or either side of the screw is not covered with product where the screw exits the 'exposed' section, then the output discharged will be reduced according to the region of screw covered. The position will also be attained where one side of the bin empties before the other and the discharge rate is correspondingly reduced.

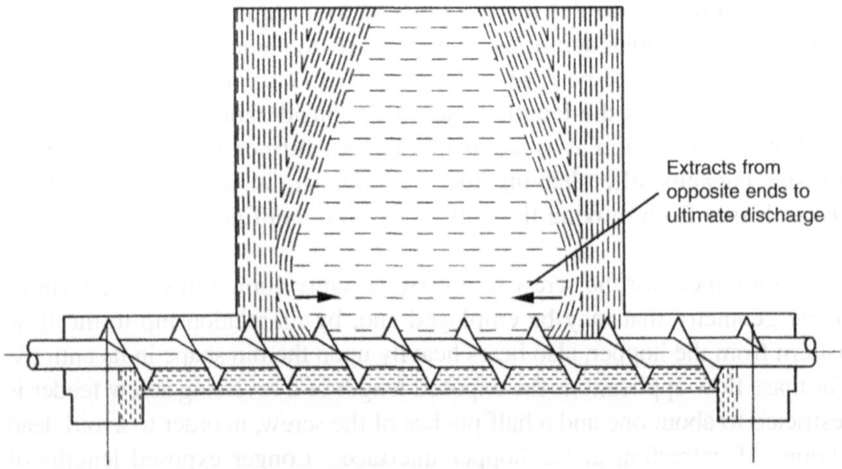

Fig. 3.11 Reversing bin discharge screw feeders

3.3.2 Screws with 'through' outlets

In some instances, discharge screws are made with an unrestricted outlet within the length of the hopper. In these cases, material can fall freely from the hopper contents, whether or not the screw is running, obstructed only by the screw shaft projecting across the opening. A separate control valve or device is necessary to stop the outflow, if required, the rate of which is not controlled by the screw. One purpose of the screw in such installations is to discharge the residue of the hopper contents, after gravity discharge has emptied the bulk of the contents. However, the screw may be used to stimulate flow by disturbing the settled condition of the material. A further reason for such an arrangement is to impose a mass flow pattern on the contents of the storage unit. Even though the flow rate is not controlled, the conversion of the whole length of the hopper bottom to a live extraction pattern enables a mass flow pattern to be generated in the hopper, albeit large velocity differences may be present over the hopper cross-section.

3.3.3 Multiple screws

Twin, triple, quadruple, and even larger multiples of screws are used in parallel to achieve large outlet sizes for storage bins. The usual reason for multiple screws is to provide the large outlet for poor-flow materials, such as filter cake and centrifuged products, sawdust, peat, and flaky or cohesive products that tend to arch over large spans. Containers with vertical side walls and screw extending over the whole base area are termed 'live bottom' bins. In extreme cases, such as the discharge of sticky

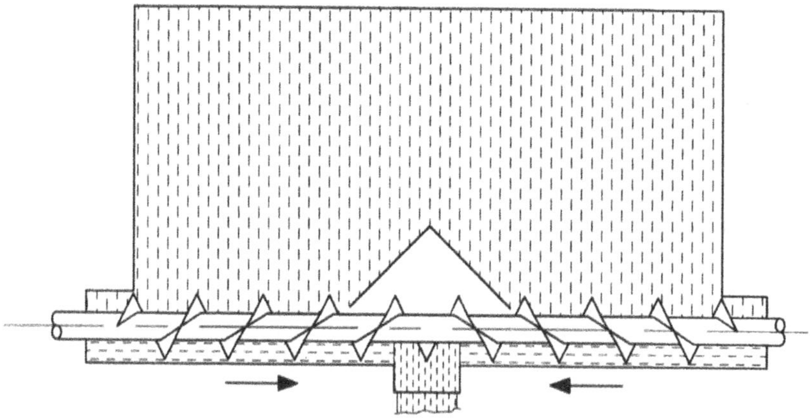

Fig. 3.12 Outlets within the length of the hopper interface

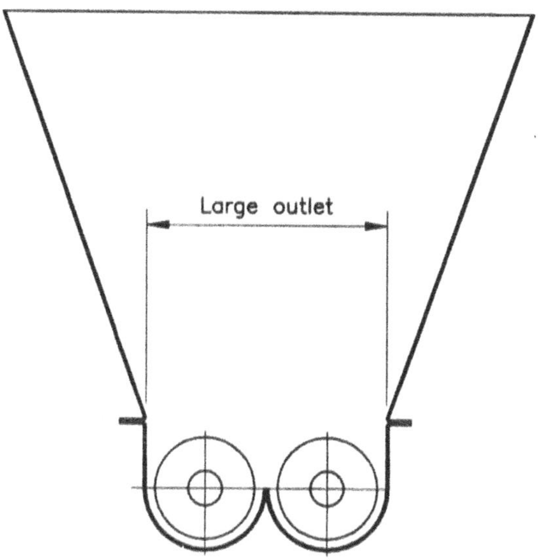

Fig. 3.13 Multiple screw bin discharge feeder

materials and those materials renown for 'bridging' and adhering to walls, negative wall slopes are employed to eliminate support for the ends of an arch. Where the width of the outlet from a multiple screw feeder is too wide for the final receiving point, a short transverse screw can be used to collect the material and focus the discharge to a single point, or to deliver to the side of the multiple screw outlet.

Drives can be individual to each screw or linked by chain or spur gears to adjacent members. It is mechanically convenient to fit such secondary drive transfers to the opposite end of the screw from the primary drive, but this arrangement requires the initial driven screw to transmit the torque absorbed by all the screws. The use of spur gears to drive adjacent screws necessitates that the screws are alternate handed in pitches to deliver to the same end of the casing. For multiple screw installations it may be useful to use two drive units. In the event of a breakdown or jam, the other unit can continue to operate and used to empty most of the bin contents. For batch weighing applications the bulk discharge may be accomplished by all screws running for the bulk of the delivery, and one set stopped for better discretion of achieving the final make-up weight, as the target setting is approached.

Large outlets allow large overpressures to develop on the screws; hence the power required to drive each screw in an array tends to be greater than that required for individual screw units.

3.4 Metering screw feeders

In contrast to the emphasis placed upon enhancing storage capacity for bin discharge screws, feeders designed to control the dispensation of loose solids have the prime objective of securing reliable, predictable, and controllable flow. This has two facets. The first is to ensure the flow channel is smooth and continuous for all conditions of use. The second is to fill the volume of the screw in a consistent manner, with loose solid material in as uniform a state of density as possible. In general, a mass flow channel best serves both objectives, for performance reliability and because the flow channel passes all the material through a similar condition of flow stresses. Mass flow also reduces the prospect of variable product condition from segregation or by dilated material forming a preferential flow channel through more stable and dense regions of static product.

In metering applications the nature of the discharge stream can also be important. The instantaneous rate of flow of loose bulk material falling from the end of a rotating screw is rarely completely uniform, even if the transport rate of material along the screw is. The three main reasons for short-term fluctuations at the discharge end are:

1. *Cyclic variations* The face of the screw flight moves material forward and there is a tendency for a gap to develop behind the working face. This is more pronounced when the pitch of the screw is expanded

to reduce material confinement after the screw has left the exposed interface region with the hopper contents. Even if the screw volume is totally filled, the rotating face of the screw at the outlet progressively uncovers the pitch volume so that a free-flowing material can surge out as the flight movement acts as the release of a reverse weir on the pitch contents. This effect is generally countered, where necessary, by avoiding low speeds of screw rotation.

2. *Cohesive failure* A cohesive material moved forward by the screw will tend to hang together, until the weight of material overhanging the outlet exceeds the strength of the cross-section of material. Erratic failures take place within each rotation of the screw as it brings forward a fresh pitch spacing of product. This effect is countered by securing a flow state of the material that avoids significant cohesive strength developing within the bulk material. Normally a mass flow channel will not bring the material into a discharge screw in a strongly consolidated state. A high speed of rotation will generate a loose bulk condition that fails more evenly. The use of vibrators and agitators to stimulate flow can cause undue product compaction to aggravate this problem.

3. *Avalanching* The fact that free-flowing materials tend to avalanche in unconfined failure is a feature that has attracted much technical attention in recent years and it is addressed by chaos theory. A slowly advancing repose slope, as formed at the outlet of a slow-speed screw feeder, tends to collapse in small surges. This effect is compounded by cyclic releases of material as the flight adjacent to the outlet rotates. Avalanching is countered by avoiding slow running speeds, or by the use of smaller-sized double or triple screws, than the single screw alternative.

The significance of short-term fluctuation of feed rate depends largely upon the nature of the application. It can have great importance where the process life in the subsequent equipment is very short, as with the feed to a mill. It may also be a disadvantage in cases where the output is visible or where the purpose of continuous feeding is sensitive to minor irregularities, as with coating a confectionery product or surface colouring a continuous product. The effect can sometimes be exploited to give texture or variety, while maintaining a consistent average rate of feed. In many instances, the lifetime of the variations is short and smoothing variations in the axial stream flow of the subsequent process attenuate most minor fluctuations. The user of a screw feeder must clarify the distinction between average feed rate uniformity, and deviations within a specific scale of scrutiny or time period, when setting out criteria for accuracy.

This overriding objective of achieving the most favourable form of flow channel for filling the screw volume with a consistent product dictates the characteristics of the screw and the form of the hopper interface. Except for feeders dealing with bulk materials that are free flowing and relatively stable in density, good feed control requires that the bulk material entering the screw should be subject to a uniform overpressure from a fully live flow channel. This generally means that the screw form is exposed over a length of between three and five diameters and has a good progression of extracting capacity along the screw axis. As metering screw feeders are commonly required to dispense relatively low output rate, the size of the screws range down to quite small diameters, in some cases 10 mm diameter or less. With this size of screw there is little scope for an extensive length of the construction to be made with progressive extraction geometry, so any 'live' hopper interface slot with a small screw feeder is bound to be short and narrow. In cases where a poor-flow material is to be handled at a low feed rate, the opening size may be smaller than the critical arching span of the material. Twin or triple screws are often employed to provide a wider opening while retaining gravity-driven mass flow benefits.

Supplementary equipment, such as agitators, vibrators, and wall-activation devices, are commonly used to promote flow. While these can overcome bridging problems and act to de-aerate fluidized powder, they introduce uncertainties about the form of flow channel which develops in the supply hopper, and the uniformity and the density of material filling the screw. Such devices can only be proven by specific use. When a mass flow pattern is needed for reasons associated with possible problems arising from extended or indefinite storage life of the material handled, then the use of feeders that do not offer a live mass flow extraction to the feed hopper contents should be avoided.

Metering screws are normally positioned to suit the application rather than designed to deal with a conveying element so, in combination with the limited exposed length of screw in the supply hopper, the overall length of the screw tends to be short. This often permits feed screws to be supported from one end. Cantilever-mounted screws require seals, bearings, or shaft obstructions only at the drive end of the machine. This feature gives great flexibility for direct end discharge, saving headroom, and allowing close coupling to process machines and the exploitation of other useful techniques described in Chapter 7. The use of hollow-shaft geared motors to provide the bearing support for the shaft also allows a compact and simple arrangement to be used for the drive mounting and thereby permits withdrawal of the screw and drive as an integrated assembly for access or cleaning.

Screw feeders are essentially volumetric devices, which in general provide uniform feed rates that can be calibrated to meet specific requirements. There are various reasons for linking screw feeders with weighing devices to offer feed control on a gravimetric basis.

1. Volumetric feeders offer no verification of the feed rate and permit no simple means to record their historic performance. Apart from an ongoing certification of performance, registered measurable values allow alarms to be actuated and failsafe or corrective action to be automatically engaged, should the feed rate fall outside acceptable limits.

2. Volumetric feeders cannot compensate for variations of material density or fill conditions of the screw. In most applications the weight of the material discharged by the feeder is the key value, whether for batch or for continuous operation. Apart from quality considerations, legislation often dictates that product weights or proportions fall within prescribed limits of stated values. Measurement of feed weight dispensed allows feedback control to be exercised through screw speed variation. Such feedback allows the feed rate to process operations to be optimized. Feed variation trends can often be detected and corrected before the consequences are apparent. As a result, higher dispensation accuracy can be achieved than with volumetric devices, which are unable to compensate for density variations of the product.

3. Electronic control systems can provide sophisticated instrumentation and monitoring based upon the registration of electrical signals, as from load cells. Typical uses are: content indication, rather than use of level controls; rapid selection from multiple recipes and variable rate delivery, as for rapid bulk discharge; and slow trimming feed to final batch weights. Control in this manner is not confined to the specific process concerned, but can be part of a computer-controlled system embracing the total manufacturing facility. This allows flexibility of individual processes, total production to be integrated, and features such as stock levels to be monitored within a comprehensive control system.

Simple batch weight control is possible by weighing the whole feeder assembly or the receiving vessel. The choice of technique usually depends upon which has the lower gross weight or offers the greater difficulty in isolating its weight from associated equipment. It is simple to re-tare the assembly and change target weights. Two speed discharge rates allow rapid bulk fill and fine discretion of the ultimate weight; these are simple to undertake from intermediate weight indication.

Continuous weigh feeding is more challenging. One method is to measure feed rate by the loss of weight of the feeder/hopper assembly and its contents. This feed process has a finite life, according to the amount of material held in the hopper, so accommodation is made for the hopper to be periodically recharged with fresh material. Loss-in-weight feeder systems typically run under feedback control for a period of about two minutes and then have stock replenished in about ten seconds. To maintain continuity the drive is 'locked' on to a fixed speed prior to the end of the weigh control period and the equipment dispenses in volumetric mode until refill is completed and a new tare weight is established. During this time, the accuracy of the feed is dependent upon selection of a screw speed that matches the target conditions of feed under gravimetric control. To avoid selecting an instantaneous speed that deviates from this, an average speed may be chosen from a series of settings taken near the end of the gravimetric period.

The decision whether to correct deviations to a target value, or an absolute average value, is application related. In cases where the overall output has to coincide with a total amount dispensed, or when subsequent flow stream attenuation will smooth minor irregularities, then correction of any shortfall or overfeed is included in feed rate adjustments. For applications where the historical feed rate has immutable consequences, then any corrections should only seek to recover from the offset to the target feed rate and not compensate for past deviations. There is no benefit in compensatory over- or underfeeding when it is not possible to rectify a prior deviation. It cannot be too strongly emphasized that flow reliability is paramount. No control system, however sophisticated, can compensate for an inconsistent feeder performance.

The high rate of flow needed to replenish the stock of a loss-in-weight feeder hopper within a short period can cause product condition problems. At the start of the make-up process material from the re-supply container commences to move from a settled state. As re-supply flow is fully mobilized, the flow stream is more dilated so the fresh product will be in a lower density than the old stock. It is important that there is not a sudden change of state of material entering the screw, as this places a high demand upon the control system. Apart from feed rate variations due to density changes, excessive dilatation of the bulk can give rise to uncontrolled 'flushing' through the feeder. It is therefore good practice to retain sufficient working stock in the feeder hopper to accommodate discharge during the volumetric period and sufficiently beyond to allow the material to settle to a consistent state of density. Features to examine are:

(i) the flow pattern that the feeder imposes on the make-up hopper (mass flow is preferred in most cases);
(ii) the size of the make-up connection (a large cross-sectional area is favourable to pass the quantity required in as dense and quiescent state as practical, noting that discharge rates of mass flow hoppers are generally lower than for non-mass flow, but flow dilatation is reduced).

Initiating and sustaining the flow of a fine powder bed is impeded by the differential between the pressure in the interstitial voidage of the particles and ambient pressure as the bulk expands to commence flow, and also as it expands further during travel along the flow path. Figure 3.14 illustrates the influence of interstitial pressure on flow through an orifice.

One way to counter the expansion restraining effect of reduced interstitial pressure is to provide a large surface area of failure for the bulk solid. The rate of permeation required to serve the reduced expansion rate is thus lowered. Techniques to promote rapid flow in this manner, without requiring gross bulk expansion rates, include:

(i) providing a large failure surface by means of a large-sized valve, while delivering through a smaller final orifice, Fig. 3.15;
(ii) flow inserts that modify the flow channel to offer a large area for confined failure, converging to a smaller channel for the higher velocity unconfined flow, Fig. 3.16.

Multiple screw feeders are also used, because they secure a large cross-section flow channel that allows high overall flowrates to be achieved without excessive local rates of expansion.

Controlled air injected can be used in two ways to encourage high flowrates from a make-up storage facility. One technique is to inject a

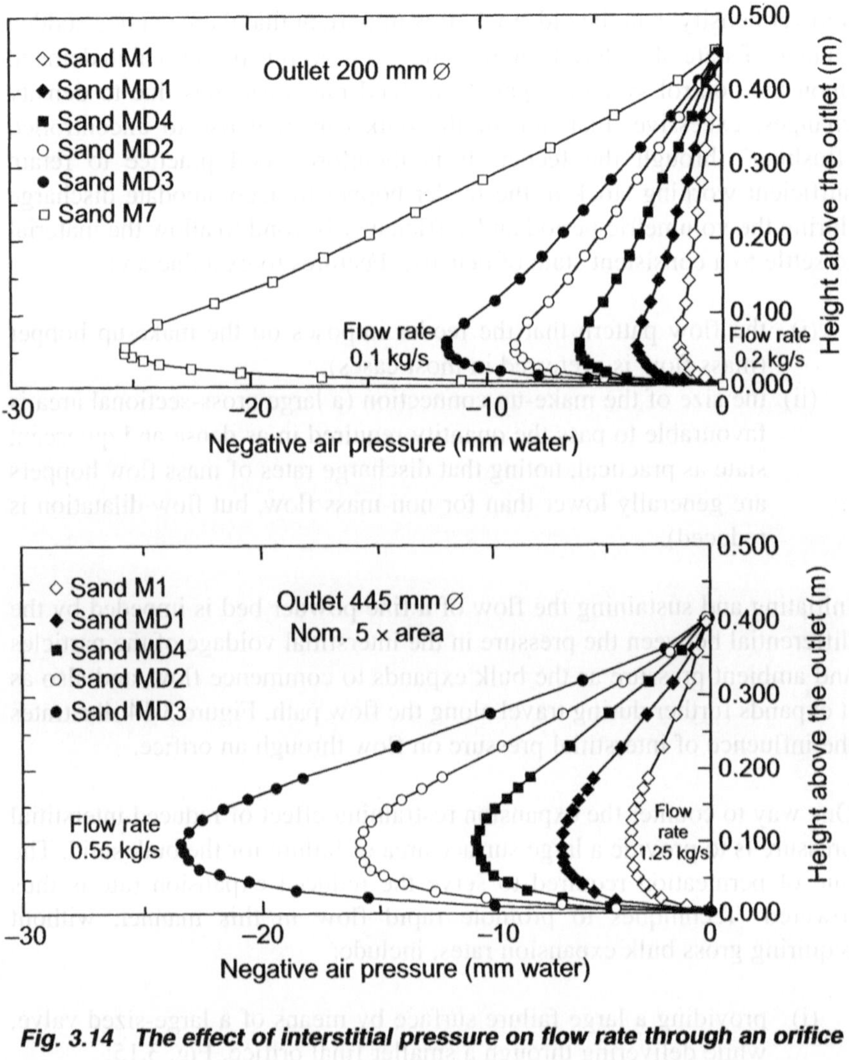

Fig. 3.14 The effect of interstitial pressure on flow rate through an orifice

small, but constant, supply of air during the whole time of storage. The purpose of this is to partly or fully replace, or even overcompensate for the amount of air diffusing from the bulk as the mass settles with time from the dilated, loose, and very free-flowing condition that it enjoyed when initially delivered into the container. Control of the powder state in this manner requires careful volume control, not a constant pressure supply that will respond inversely to the resistance. This method allows the re-supply stock to be held in a condition that will flow readily and swiftly when the release valve is opened.

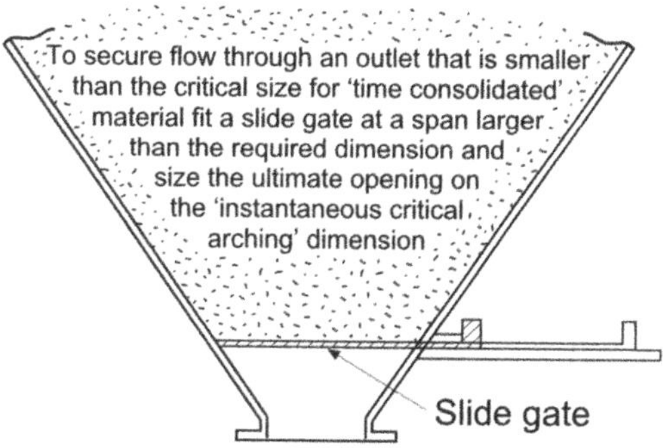

Fig. 3.15 Large valve to initiate flow

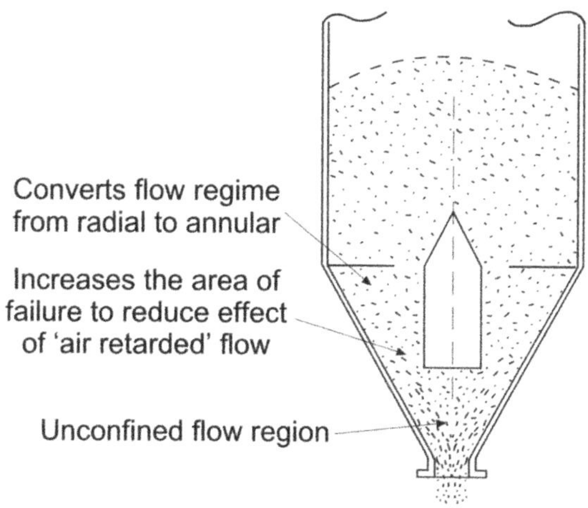

Fig. 3.16 Insert to provide large flow area

Another way of stimulating flow and satisfying the void expansion demands of the bulk as it starts and continues to flow, is to inject air into the bulk above the outlet as the delivery valve is opened. This supply may be steady, or in the form of an initial small surge reducing to a lower rate after flow is initiated. In either case, the amount of air should be calculated and controlled to suit the particular circumstances of the application. In cases of continuous air injection the effect of possible contaminants, such as moisture, must be avoided where it would have a deleterious effect upon the condition of the bulk material. In all cases, attention must be focused upon the condition of the product as received by the feeder hopper, and how this condition is stabilized to a consistent, controllable state prior to entrainment by the feeder screw.

Chapter 4

Screw Construction

4.1 Mechanics of screws

An essential feature of the working of a screw feeder is that the whole cross-section of the screw is filled with bulk material, replenishing material transferred along the screw axis by further inflow from the supply hopper. As a consequence, in contrast to a screw conveyor that operates with a partially full cross-section, gravity has no bearing upon the motion of the contents. The weight of material filling the 'rising' side of the rotating screw feeder is balanced by the weight of material occupying the 'descending' side. The quantity of material displaced is determined by the screw geometry where the screw exits from the exposed section of the hopper inlet, even if the screw is filled at some prior location along its axis. The pressures provided by the constraints of the casing and overpressure of material from the hopper determine the compacting stresses on the conveyed contents, and hence the density that the material attains in transit.

The other two factors influencing the feed rate are the volume moved forward by the screw, and the effective axial transfer velocity of the material. Unless there is a degree of residue adhering to the screw blade, the volume moved is bounded by the screw geometry and the outer failure surface of the moving material. The outer surface of the moving material does not necessarily correspond to the outside diameter of the screw but can take various boundary shapes. Examination of the detailed mechanics of the screw behaviour requires an analysis of the screw contact face mechanics, and an understanding of bulk rheology.

Bulk material is promoted to move in a screw feeder by rotation of the inclined face of the screw, acting as a travelling wedge on material within its swept volume. Sliding must take place on this working face for the material to move axially; friction of the material on the flight surface must be less than the internal friction of the bulk material otherwise the material would rotate with the screw. The direction in which the material moves is dependent upon the frictional co-efficient on the face of the flight. Boundary friction on the outside of the moving mass does not influence the direction of motion of the bulk material around the screw axis. It does, however, determine the degree of frictional resistance opposing movement, and hence the power needed to turn the screw. If the particles were free to travel an independent path, their relative displacement would follow a helical path at an inclination related to the local helix angle of the blade and the angle of contact friction on the flight surface, see Fig. 4.1.

The blade helix angle varies continuously according to the radial location. This is a consequence of the peripheral distance travelled by any point of

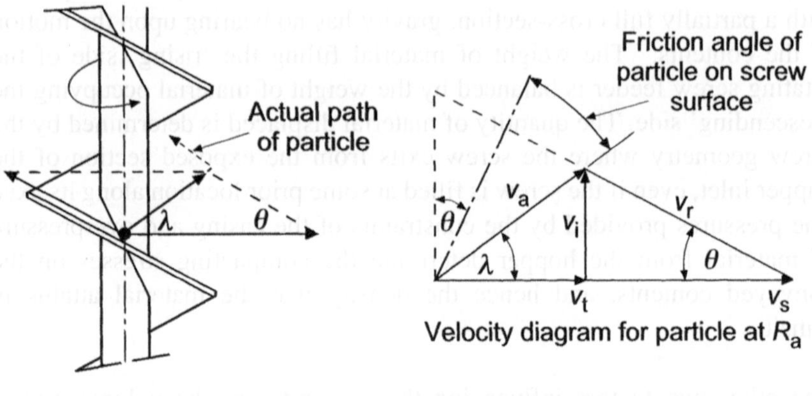

Fig. 4.1 Path taken by an independent particle sliding on the flight surface

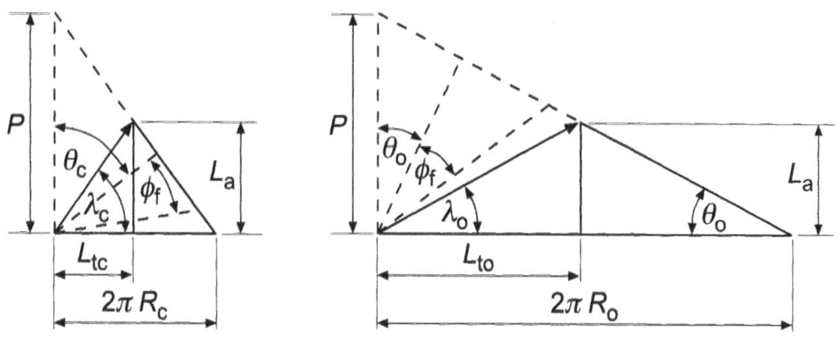

Fig. 4.2 Variation of flight helix angle with radius (screw diameter, D = 2R_o; shaft diameter, d = 2R_c)

location being dependent upon the radius, while the pitch of the screw is constant for all radial positions. The angle of inclination of the screw flight therefore varies from $\tan^{-1}.\pi(D/P)$ to $\tan^{-1}.\pi(d/P)$, see Fig. 4.2 (D and d being respectively the diameter of the screw and its centre shaft).

Two features determine the axial progress of the material occupying the swept volume of a feeder screw.

1. How the material in contact with the face of the screw flight moves with respect to its radial location on the face.
2. How the non-contact material at this radius moves with relation to the material on the contact face.

The shear strength of the bulk material in transit determines whether there are differential displacements between differing regions of the conveyed volume, according to the radial or axial location in relation to the flight face. The boundary conditions are crucial, as material not in contact with a boundary is impressed to move with the surrounding material, no other forces being present.

Considering the cross-section of material within one pitch space of the screw, the relevant containment boundaries are:

- the back or non-working face of the screw flight;
- the inner shaft periphery;

- the working face of the screw flight;
- the outer periphery of the feeder casing or static residue on the under side and sides of the outer diameter; and
- the in-flowing product, or the static head of material resting on material over the upper region of the screw's periphery.

Regarding the motion of material in contact with the 'working' face of the screw flight, the two limiting conditions are that the material moves as a coherent mass, or that regions move along differing helixes, according to the local radial and axial conditions.

For coherent movement of the screw contents, the common helical direction taken by the bulk will follow some mean value. This value is dependent upon conditions at the respective areas in contact, and on the local contact pressures. The geometry and location of the restraining shear that prevents the material from rotating, heavily weights the effectiveness of the outer regions of the screw flight surface, in terms of area and contact pressure.

A 'mean' path of material motion, based on area alone, indicates the dominant role of the outer regions of the screw flight (Fig. 4.3). This value is a conservative basis on which to calculate the screw transfer rate. The higher surface contact pressure on the outer radial surface of the screw flight compounds the effect of the more favourable conveying angle over this area. The helix angle of the screw, θ, varies from θ_o at the outside tip of the flight at R_o to θ_c at adjacent to the centre shaft, R_c. The mean helix angle θ_m is at R_m, where the area of flight face between θ_o and θ_m equals that between θ_m and θ_c. ϕ_f is the angle of friction of the bulk material being handled against the face of the screw flight. λ is $90^o - (\theta + \phi_f)$ at the various radii; L_a is the axial advance of particle per revolution of the screw; L_{tm} is the circumferential movement of particle per revolution if the screw at 'mean' diameter; K is the ratio screw diameter: shaft diameter; P is the pitch of the screw; and λ is the angle of spiral movement of a particle.

Calculations based upon integrating the movement of independent concentric rings of material with internal shear, so that the motion of annular sections follows a theoretical path according to the specific helix angle at particular radii, give very similar predictions of capacity.

There is, however, a further complication arising from centre shaft contact with standard forms of feeder screw construction. Rotation of the screw shaft makes no contribution to moving the material forward; in fact it

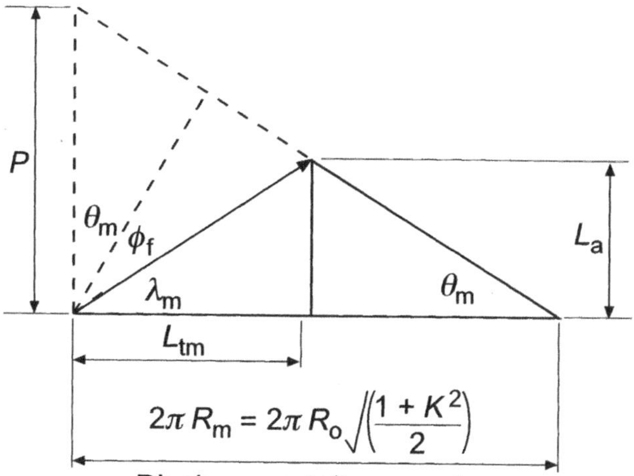

Displacement diagram at 'mean' radius

Fig. 4.3 'Mean' path of motion for material in contact with the flight surface

offers a frictional restraint to forward progress. The 'corner effect' between the shaft and the flight face also resists clean slip. Shaft effects, other than corner build-up, are relatively minor. The contact pressure on the shaft is low, the surface is generally smooth, and the area of contact is small by comparison with both the areas of the screw flight face and the outer boundary periphery. Nevertheless, large-diameter shafts do have an adverse effect, particularly when cohesive or damp materials are handled. These corner 'fillets' of unmoving material adhere to the shaft and the inner region of the flight face. The 'trailing' face surface of a screw flight is more sheltered than those exposed to working face pressures. Adhered residue therefore invariably forms small radii on the working face corner, and larger radii on the back, or 'trailing' face of the flight, where less force is available to dislodge material. A typical profile of residual 'build-up' is shown in Fig. 4.4.

Material carried within the cross-section of the feed screw is deterred from rotating with the blade by the boundary conditions at the outside radius of the screw flight. Frictional resistance to rotation is offered over the area of the screw periphery that is not exposed to the inlet flow from the hopper, by material in contact with the outer casing surface. This restraining surface is radially displaced from the tip of the screw flight, by the working clearance of the screw in the casing. Products which 'shear thin' can tend to 'log' in the screw due to the low restraint offered to

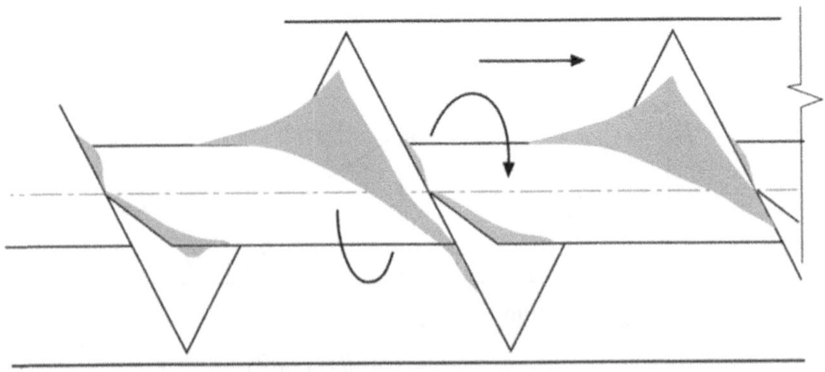

Fig. 4.4 Typical build-up of residue in flight/shaft corners

rotation of the mass. Alternatively, a smooth, circular boundary surface of low contact friction can allow the material in the residue layer to rotate with the screw, and similarly negate forward motion of the contents by 'logging'.

The behaviour of material in the boundary layer between the flight tips and the casing is significant to screw transfer capacity. This clearance dimension tends to be chosen from such considerations as fabrication tolerances, avoiding casing contact by screw deflection, minimizing residue, and avoiding particle trapping, rather than any effect upon the efficiency of transfer capacity or power absorption. The effect of movement in the clearance layer is proportionally much higher for small screws than larger diameter units. For example a 5 mm wide tip clearance on a 50 mm diameter screw with a 25 mm diameter centre shaft, has an annular clearance area equal to 92 percent of the screw cross-section. By contrast, the annular clearance area on a screw 400 mm in diameter with a 75 mm diameter shaft and 10 mm tip clearance, has less than 15 percent extra area in the annular cross-section. Feeder rate predictions or screw efficiency calculations that do not include an assessment of possible boundary layer movement can be misleading, and 'screw efficiency' calculations are meaningless.

In practice, the material in the boundary layer surrounding the screw may behave in any of three ways, according to the nature of the material and its relationship with the various contact surfaces. The whole boundary layer may move with the contents of the screw, to wipe or scour the inner surface of the casing walls. This conduct is likely with fibrous products, dendretic particulates, materials that have a large granular nature or high shear strength, combined with low friction properties on the casing surface. Material that moves in this manner tends to advance as a coherent mass, and volumetric output rates are as if the screw diameter extended to the wall surface.

Fine, but free-flowing materials such as small rounded grains, metal powders, and non-caking minerals, will rest in the lower clearance layer. Product resting over the top surface of the screw contents will travel on the moving bed in the screw, which is advancing in a spiral path. This pattern of behaviour results in an eccentric cross-section of the material in motion. As a rough guide, approximately 25 percent of the boundary layer can be considered to move with the screw. In some situations the material in the lower boundary layer may be disturbed, to creep or erratically move in an intermittent fashion.

Fine cohesive powders, and materials of high shear strength, tend to form a static annular layer around the contents of the screw. This layer can stay in place when the screw itself runs out of material, although the more usual result is for the top section to collapse and discharge when voids appear in the screw volume as the feed hopper empties. A strong permanent bed around the screw tends to become firmer as particles trapped by the flight tip further compact the bed, or as it hardens due to various 'caking' processes. The absence of an effective tip clearance then gives rise to high torsional loads and wear of the outer rim of the flights.

A potential source of error arises when assessing screw transfer rates based upon values of bulk density in static conditions. There are two aspects: the proportion of swept volume actually occupied by material; and the true density of the material as conveyed. Regardless of the filling pattern there is always a partial void behind the working face of the flight as it moves away from the material. This is usually small unless the screw is running at high speed. The density that the material attains is very pertinent to the output rate in mass terms. Products that exhibit small variations of density, such as compositions of fine crystals or granules, are normally more dilated in transit than when in a settled state due to shear dilatation. A simple way to assess this class of materials is to measure the densities as 'loose poured', then as 'lightly tapped', and extrapolate back an equivalent amount, to compensate for the effect of the material's dynamic behaviour.

Powders that exhibit a wide range of density states are invariably sensitive to prior formation conditions, i.e. their stress history and flow channel life. The most useful approach is to ensure that: the route prior to entering the screw is of mass flow nature; sufficient time is available for the material to achieve a stable condition; and the flow area is adequate to pass the rate of feed without serious inhibition from the void expansion demand. Design attention directed to achieving consistency of density is rarely misapplied. One of the advantages of a screw feeder is that the flow area over the inlet length of the screw provides a large cross-section for ingress of air, to satisfy the smaller rates of void expansion than would be needed for similar flow rates through a small flow orifice. For materials with highly variable densities, measured values are best secured by replicating the flow conditions in a similar manner to the installation under consideration.

4.2 Screw forms

Conventional feeder screws have full-face flights welded to a centre shaft or tube. Uniform pitch screws are sometimes used in simple feeding applications. In order to secure differential intake of product along the axis of a screw, a range of techniques can be adopted, either individually or in combination where appropriate. Typical constructional features are (see Fig. 4.5): stepped pitch; variable pitch; stepped or taper centre tube or shaft; variable screw diameter; part ribbon or shaftless construction. (Occasionally the construction is stepped on the outside diameter.)

4.2.1 Flight construction

Flights are made in a range of thicknesses. Thicker flights are normally specified for either strength or wear resistance, although the intrinsic shape of the blade forms a stiff construction and wear, when it is a factor, is more usually on the tip, rather than the face, of the blade. Abrasive resistance steels are used for applications handling rough and aggressive materials. Hard weld deposits are applied to flight rims or outer regions of the working face of the flights, to deal with very abrasive products. Blade thickness has only a minor influence on the transfer capacity, as the width occupies a small portion of the swept volume. This is rarely significant except in the case of small-diameter, short-pitch flights, where the blade thickness can be a significant proportion of the flight pitch. Thick blades exacerbate 'tip drag' when the residual layer of product offers a firm bed and particles trapped under the flight tips bind to create high pressures and large frictional forces.

To resist face wear, radial ridges of hard weld deposit, applied at short intervals, trap pockets of fines to form thin 'wear boxes' on the flight face.

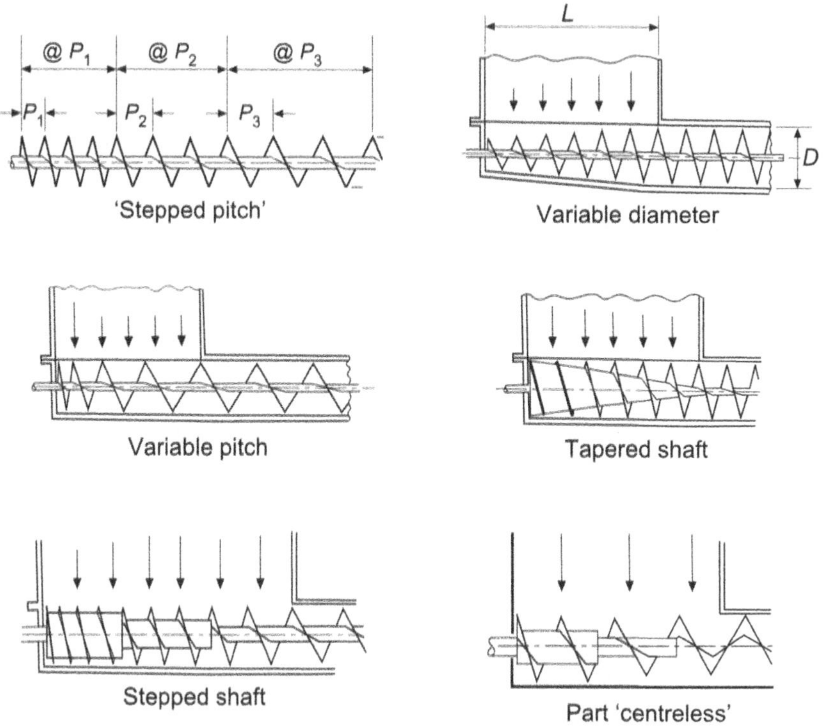

Fig. 4.5 Forms of feeder screw

A useful technique to reduce tip drag and the confinement zone area for tip binding, is to chamfer wide flight tips to a thin contact width on the working edge. The angle of chamfer can be coarse, so long as the lagging boundary runs clear of the swept surface. This approach is also used to reduce or avoid shaft 'juddering', as when a slender screw cyclically bounces on a firm residue bed of product, such as a damp filter cake. The sharp flight tip will scrape the surface of the bed, rather than chatter and bear upon the surface without the pressure to penetrate.

Flights formed from a continuous strip are economical to form in quantity. The process essentially deforms the cross-section of the metal by thinning one edge of the strip from its original thickness. Individual pressed flights, by contrast, are fabricated by twisting an annular ring by means of a suitable press tool. A suitably proportioned blank disc can be pressed without change of the radial width or stretching of the inner and outer diameters, thereby minimizing the inherent stresses in the final flight form. The formed flights cover more than 360 degrees of the shaft on to which they fit. Therefore, the flights can be 'snapped' on to the shaft as a close-fit, stable component to match the adjoining member for butt-welding of the contact edges. Operating efficiency is dependent upon accurate alignment and smooth transfer surfaces between the flights. Small-diameter flights are particularly sensitive to joint smoothness, as weld widths and finishes are independent of the size of the flights.

4.2.2 Centre shafts

Solid centre shafts are heavy relative to their strength, compared with tubes; hence solid shafts tend to be used only on short feed screws. Bolted-in end shafts, a fairly common form of construction for some standard ranges of screw conveyors, are rarely used for feeder screws, as these are less strong than welded-in end shafts and present hazards relating to hygiene, cross-contamination, crevice corrosion, and fatigue failure.

Ribbon screws, and screws without a centre shaft, have advantages for handling cohesive, wet, and sticky materials. This screw construction substantially reduces the tendency for bulk material to adhere to the central region and shaft of the screw. Specialist forms of ribbon construction have been developed to further reduce the tendency for shaft build-up. Ribbon screws with 'crevice-free' welding to centre shafts are also employed to reduce the prospects of fatigue in stainless steel screws. This hazard afflicts machines that work continuously and subjects their shafts to large numbers of repetitive cycles of rotary bending during their working life. Special attention to welding and finish should be given to stainless steel screws because of the susceptibility of many grades of stainless steel to fatigue failure.

Provided that the remaining width of the screw flight is adequate, ribbon and 'shaftless' screws transfer more material in flood-feed conditions than full-bladed flights. The reason being that the material in the central region of the flights moves coherently with the surrounding product, and the inhibiting effect of the central, coarse helix and corner connection with the centre shaft are eliminated.

In some cases, individual paddle flights are used on feeder shafts. Their main virtues are:

- they provide a degree of breakage or mixing of the product handled;
- the blades can be adjusted to allow some control of the extraction pattern;
- new blades can be fitted if the originals are worn or damaged.

The greatest degree of flight 'openness' for lump breaking duties is given when individual paddles are mounted in a contra-helix pattern along the shaft. That is, the points of mounting into the central shaft follow a helical path of opposite hand to the inclination that forms the handling of the screw form. Special forms of blade profile and support rib arrangements have been developed to maximize the cutting and breaking facility, and also to minimize the tendency for material to build up around the central shaft region.

4.3 Materials of construction and finish

Screw feeders are normally of metallic construction because of the cost/strength requirements for industrial applications. Some small feeders are constructed of non-ferrous materials to allow metal detectors to be used around the casing, rather than on the free fall of the flow stream. There are also casings made of flexible materials, for vibrators or oscillating pads to deform the walls in order to stimulate flow. While these are usually successful in promoting flow, the regime of flow that is created is generally uncertain and in most cases definitely not of a mass flow nature.

Mild steel is commonly used for large-scale feeders, for handling minerals and similar bulk materials that do not present a corrosion or hygiene problem. Most units in mild steel are of natural finish on the contact surfaces, with external surfaces painted. Alternative surface finishes for protection include zinc spray and sometimes galvanizing – although galvanizing can be a very hazardous treatment to undertake, because of the distortion that may occur in the dipping process. Relatively thin sheet metal fabrications, such as the hoppers and casings of feeders, are prone to severe distortion during galvanizing. Buckled or twisted casings play havoc with design clearances and can prove awkward to straighten. Even more difficult to correct are deformed screw augers, because it is not practical to utilize the normal heat treatment means to rectify bent shafts and tubes.

In the case of small screw feeders the cost of the material of construction is of low significance to the overall manufacturing cost. Stainless steel construction offers many benefits for universal application. Its main virtue is that the contact surface is suitable for many materials, as it resists corrosion and deterioration with time due to adverse ambient conditions, and it offers a hygienic finish. The equipment can be hosed down or steamed out for cleaning, sterilizing, or decontaminating, without problems of rust. This is an important feature with respect to sustaining consistent slip characteristics on the contact surface of the feeder hopper and the screw flights. For predictable, reliable flow, and for consistent feed performance, it is essential to have stable, and preferably low, contact friction conditions. As a consequence of these many demands, set against the marginal cost of fabrication, stainless steel construction is widely used for metering screw applications.

4.3.1 Surface finish of stainless steel

The specification of a surface finish for stainless steel equipment is a source of many potential problems and misunderstandings between purchasers and manufacturers. The implications for costs and like-for-like comparisons between tenders, and the need to establish that the equipment condition satisfies all functional requirements of the application, dictate that the surface finish specified should be appropriately defined in all contractual documents. Descriptions such as 'crevice-free' or 'polished 400 grit', may appear definitive, but all too often lead to ambiguities, disagreements of interpretation, or unfitness for purpose.

The selection of surface finish is usually made on the basis of choosing the optimum economic method to meet the objectives related to suitability for the application. The choice may be compromised by global conditions, such as seeking machine or plant uniformity, or by narrow key functions of crucial importance, as with fatigue problems or stringent duty constraints.

A useful starting point is to review the differing criteria on which the specification of surface finish is based. This illustrates the 'many attributes' nature of the situation and how these may apply individually or in varied combinations, according to the application circumstances. The reason for the selection of a particular type of surface finish usually relates to one or more of *four basic considerations,* relating to:

- appearance;
- performance;
- product; and/or
- plant.

4.3.2 Appearance

The most common reasons for preparing a machine to a particular standard of appearance are given below:

Cosmetic A cosmetic finish is given for superficial appearance, for decoration, or to conceal blemishes. Highly polished parts give a superficial impression of quality and purity, emphasizing the nature of the material of construction, but do not necessarily represent the implied standard of detail finish. Bright, polished panels, or covers that are easy to fabricate without scratching, enhance the presentation of an otherwise uninteresting product. Unsightly manufacturing and constructional marks are removed to prevent the user viewing the item as a crude fabrication. It is common to acid clean and wire brush welded parts, to remove weld discolouration, and to blend in softened and de-scaled (S & D) or 2B finish surfaces. It must be recognized that the finish of a cold-rolled 2B stainless steel surface cannot be replicated by mechanical polishing. A separate agreed specification must therefore be defined for welded parts of assemblies that are categorized as 2B finish. Likewise, the treatment of scratches or other surface blemishes should be predetermined. One approach, to minimize minor damage during handling and fabrication, is to apply a plastic coating to the mill sheets prior to delivery to the equipment manufacturer. This protective film can be removed at a late stage of equipment construction. To some extent, this coating inhibits cutting and welding, as laser or plasma profiling of the covered sheets causes edge fusion of the plastic, and an extra edge dressing process is then required in order to secure a clean weld.

Uniformity Where a machine or component is to form part of an integral assembly, it is usual to blend all sections of the composite equipment to a common appearance. This essentially requires visual alignment of the surface textures and finish, to avoid detraction from an overall appearance of quality. Generally this requires the concealment of manufacturing scratches and blemishes, weld and burning discolouration, and planishing marks. Bright, or mirror-finish, sections may be utilize, to command attention and draw the eye from constructional details and edges, which may be left unwelded to avoid distortion and discolouration.

Conformity Plant consistency, in line with the user company's policy or image, is served by new equipment having a similar grade of appearance to that previously installed. This matters in situations where the perception of quality, purity, and hygiene, are important to the reputation of the user, for example in dairies, food factories, and pharmaceutical plants.

Manufacturing Standard A standard of finish may be set by the manufacturing company, or by the user. Manufacturers offering equipment to industries that expect a particular standard of finish may implement a policy of preparing to a set finish, regardless of the use to which the equipment is finally put. A feeder manufacturer offering a standard machine to have universal appeal may decide to adopt a surface finish that meets a range of application objectives. The pleasing appearance not only reflects the quality of construction, but also is able to satisfy an 'easi-clean' and hygienic finish, corrosion and stress resistance, non-contamination, and good surface slip properties within a common standard, although some of these attributes may not be essential on every application.

4.3.3 Performance

Surface slip Hoppers, chutes, and screw feeder blades all require relative motion to occur between the contact surface and the material handled. The performance of a feeder depends upon the equipment mechanics, and surface friction is a key design parameter for screw feeders and supply hoppers. Surface slip determines the effectiveness of the flights in moving the material, and thereby influences the extraction pattern from the supply hopper. The design procedure for determining the wall inclination needed to secure mass flow in conical and wedge-shaped hoppers is based mainly upon wall friction tests. For hoppers that are not of mass flow type, wall friction is important in relation to the slope at which the material will slide clear of the walls to empty the hopper. Frictional values are fundamental to the performance of any bulk handling equipment, and so simple to undertake that these measurements should be an integral feature of any feeder specification, or indeed any solids handling contract.

The flow regime in a feeder hopper, and pressures acting at the screw interface, depend upon whether the material slips on the walls. Wall friction for a given surface finish must relate to a specific bulk product, because there is no correlation between differing bulk materials for surface slip behaviour with respect to a given finish. For example, 2B finish may offer superior sliding properties to an electropolished surface with one material, but higher resistance for a different product, or even for the same product with differing moisture content. Quantified values are essential and the properties of the bulk material must be specified with precision.

Surface release properties A smooth surface tends to discourage the adhesion of cohesive, damp, wet, or fatty solids, although when the avoidance of such behaviour is important, the surface cohesion of optional surfaces should be measured by tests. Surface cohesion is shown on a graph, whereby the force promoting slip is plotted versus the normal load. The intercept on the slip force axis shows the measure of surface cohesion. A bright, polished surface is not normally required for surface cleanliness or to prevent the adhesion of particles, in fact it can be adverse. However, the marginal cost of securing a bright polish is not high, hence surface finish may be enhanced for cosmetic effect or other reasons, at relatively little cost.

4.3.4 Product

'Easi-clean' Feeders that dispense differing or incompatible products, as with different coloured pigments, flavours and spices, or pure chemicals, need to be cleaned between applications, either regularly or on a campaign basis. For ease of cleaning the surface must be smooth and with an absence of cracks or crevices. Continuous welding of butt, angle, and 'tee' joints may be adequate for some applications, but when the parts have to be scrupulously cleaned to avoid cross contamination, again a 'crevice-free' finish may be specified. Apart from the ease of cleaning, there is the need, in many instances, to verify that there is no prospect of cross contamination. Visual inspection is simplified when the absence of surface irregularities permits easy confirmation of the thoroughness of the cleaning process. In all such applications the machine design obviously plays a large role in providing access for the cleaning process, but there may be a trade-off between the standard of finish chosen for the equipment, and the need for more thorough access or even having to dismantle the equipment for cleaning.

Sanitary Hygiene criteria apply in many food-handling and pharmaceutical situations, where the growth of bacteria or fungi in cracks, crevices, and inclusions is a serious hazard to health. The relevant scale and effects are not visible to the naked eye, hence a testing process for cracks, such as dye penetration, is essential to verify the standard of surface finish. There are no great problems in achieving a hygienic surface on a smooth and continuous sheet surface. Greater difficulties arise with welded butt joints, and more so with concave and square welded joints. As a minimum the welds should be continuous. Argon welds are sometimes used to present a well-penetrated, if rippled, surface. For many screw feeder applications handling foodstuffs or organic materials, the continuous welding of all parts that are in contact with the product provides adequate hygiene standards. Regular cleaning, washing down, or sterilizing, in these circumstances, can suffice for normal operation. Sterilized or washed components may require drying before re-use, and a smooth surface offers an easier surface to dry.

For more stringent duties, such as handling fish, meats, and dairy products, and where there are extended periods of operation between cleaning, the welds should be ground and polished 'crevice-free'. The unqualified definition of 'crevice free' is, to some extent, ambiguous and subject to many interpretations. However, for sanitary purposes 'crevice-free' welds should be ground to a smooth corner fillet, or polished to form a continuous smooth surface of the same standard as adjoining metal. There should be no cracks, crevices, inclusions, weld splatter, or irregularities that show when subjected to dye penetration testing. A degree of undercutting from the dressed weld into the parent metal is almost inevitable, but permissible, so long as the surface is smooth. This is a reason why this finish is not compatible with thin metal construction.

The specification of a 'crevice-free' finish does not necessarily imply a highly polished surface. The surface finish, in terms of fineness of grit used, is a separate issue to the absence of crevices, and this should be clearly brought out in any specification of this nature.

4.3.5 Plant – machine integrity

Wear resistance The casing of a screw feeder is not normally exposed to wear from the product sliding on the surface because the clearance between the screw flights and the casing usually allows a dead layer of product to isolate the inner casing surface. Flight tip wear is the most common form of wear afflicting screw feeders. As previously described, thicker flight forms do not reduce flight tip wear. In fact the thickness detracts from an extended life, because any product trapped between a thick flight tip and the bed of residue takes longer to clear as the screw rotates than it would from a thin flight. Counter to intuition, a flight chamfered on the trailing edge reduces tip wear and has the extra advantage of reducing the power needed to turn the screw. A narrow, hard weld deposit on the tip provides the same effect, and is a good insurance for abrasive duties. Face wear is not usually a problem, except with very abrasive materials.

Austenitic stainless steels cannot be hardened by heat treatment but are strengthened by cold working, retaining good ductility and toughness even at high strength. Shot peening of the working face may therefore be used to harden the surface. The shallow surface 'skin' that is compacted in this manner induces a compressive stress in the boundary layer of the material. When the material is loaded under tensile stress, the initial strain serves to reduce this compressive stress. It is only when the initial condition passes through equilibrium that the surface experiences an effective tensile stress. As tensile stress is the main culprit in stimulating fatigue failure, the resulting reduction in value of the absolute tensile stress, when a feeder shaft treated in this way rotates and suffers rotary bending, offers considerable extra effective strength to oppose the onset of fatigue failure.

For aggressive wear applications, a wear-resistant deposit can be welded or flame applied to the flight face. One technique analogous to 'wear boxes' on chutes is to lay spaced radial runs of hard weld deposit, to capture 'fines' in the spaced pockets. A thin bed of the material being handled then presents a layer of like material as a working face. However, the surface slip behaviour on the screw face is then a factor of the internal friction of the material, rather than of a smooth metallic interface.

4.3.6 Corrosion

This can have many forms, for which different finishes are appropriate.

Pitting corrosion Austenitic stainless steels owe their corrosion resistance to the formation, from the chrome (Cr) content of the steel, of a passive layer of chromium oxide on the exposed surface. Pitting occurs when the protective oxide film breaks down in small, isolated spots. The rate of attack tends to increase because of the differences in electric potential between the large surrounding passive surface and the active pit. This action is accentuated by the presence of saline solutions. A smooth surface, free of sensitive local minute pits or small depressions, reduces the potential for pitting to commence. The most appropriate quality of stainless for such duties should be selected.

The molybdenum (Mo) content of types 316 and 317 stainless resists the onset and development of pitting. The presence of nitrogen (N), a strong austenitic former, remarkably increases resistance to pitting and crevice corrosion. The relevant effect of Cr, Mo, and N on crevice corrosion is denoted by the widely used Pitting Resistance Equivalent Number (PREN).

$$\text{PREN} = \text{Cr}\,\% + 3.3\,\text{Mo}\,\% + 16\,\text{N}\,\%$$

This indicates that the resistance of a stainless steel to pitting corrosion is increased as the proportion of these alloying elements is increased, but the influence of N is 16 times greater than the Cr content and Mo 3.3 times greater. Typical values of common stainless steels are given in Table 4.1.

Crevice corrosion While the quality of the stainless steel is the prime defence against corrosion, the presence of cracks and crevices allows

Table 4.1

Spec.	PREN	Cr	Ni	Mo	C max.	N max.
304L	18–20	18–20	8–10.5	0	0.03	0.1
316L	28	16–18	10–14	2–3	0.03	0.1
316NL	205–240			0.03–0.08	0.16	

stagnant residual pockets of aggressive substances to remain. These imperfections shelter local differences in oxygen concentrations, associated with deposits on the metal surfaces that cause progressive damage, after superficial cleaning of the outside surfaces. The ambient conditions pertaining within a sheltered crevice may be totally different to unconfined surface conditions, allowing more aggressive corrosion conditions to be developed and sustained than apply to the outer surface. The propagation of cracks can be very rapid; this is especially intense in chlorine environments. Molybdenum-bearing steels are used to minimize this problem. Cracks and crevices also present large surface areas for corrosion attack and may lead to mechanical weakness in the component, should the local region suffer excess corrosion. The specification 'crevice-free' finish may be called for to avoid such hazard, but the smoothness of the surface is not so crucial as for hygienic duties.

Stress corrosion cracking The above situation is greatly aggravated in circumstances where accelerated corrosion takes place under a combination of high tensile stress and a corrosive environment. The presence of a crack invariably leads to points of stress amplification, and chloride environments are extremely aggressive in conditions of high stress. For these situations, a high degree of surface integrity at a fine scale is essential. The grinding and polishing of corner welds, to blend in with the complete absence of stress raising conditions, is an essential step to counter the aggravation of stress corrosion attack. As with the fatigue problems described below, steps should be taken to reduce tensile stresses where practical. Type 317 with a minimum of 3.5 percent Mo is recommended for such duties.

Galvanitic corrosion A battery effect is created when dissimilar metals are in contact in the presence of an electrolyte. The flow of current causes one of the materials to corrode preferentially. In these conditions, any metal in the galvanitic series tends to suffer corrosion when in contact with a metal of more positive potential in the emf series. The other metal in contact will experience little or no corrosion. Where this is relevant, it is important to know which material is most anodic (least noble). The composition of weld rods, bolts, other fastening devices, and adjoining materials, must be compatible with the parent or contact material in all situations where there is any potential for galvanitic corrosion.

Fatigue failure Most stainless steels do not have the fatigue resisting qualities of many other types of steel. They are therefore vulnerable to cyclic and repetitive stress applications. Screw feeders in continuous operation can soon accumulate high numbers of stress reversals in the screw shaft, as a result of rotary bending, to bring the component into the range of fatigue sensitivity. The first line of defence is to ensure that the level of normal stress is relatively low, by specification of a suitable tube or centre shaft size for the span and torque to be accommodated. The second stage is to reduce stress amplification points. The root of any tiny crack or surface irregularity provides an effective stress concentration and offers prime sources for the propagation of fatigue cracks. The surface condition of the screw is therefore of major importance, and should be free from sharp included corners, cracks, and imperfections. However, welded joints and connections that do not have full weld penetration shelter internal discontinuities similar to cracks, often little removed from the metal's surface. These also provide a weak region and stress concentrator, to create a prime source for the propagation of a fatigue crack. The specification 'crevice free' in this case is not just relevant to the exposed or contact surface, but should be taken to affect the general construction with respect to internal flaws and cracks. These are only verifiable by radiography of the weld regions.

The use of a ribbon construction for the feeder flight significantly reduces the run of shaft exposed to the formation of cracks; therefore, in sensitive cases this may be considered as a constructional method, other factors being compatible.

Shot-peening is used to alter the surface condition of stainless steel subjected to fatigue conditions. Essentially the initial imposition of extra surface stresses causes local deformation that re-adjusts the stress state within the region. This process relaxes high local stresses. The subsequent creation of a thin layer of compacted steel introduces a residual compressive surface stress. Under usage conditions, externally applied tensile stresses are offset by this inherent compressive stress, as a corresponding reduction of the maximum tensile stress level is achieved. A characteristic of fatigue behaviour is that a small reduction in effective stress value at the crucial failure levels will transform the life potential of the component, in many cases extending the magnitude of the cyclic stress capacity to a useful or indefinite life. Tensile stress capacity is almost invariably the most damaging form of stress, so this technique of superimposing a compressive stress on the surface layers of components allows higher stresses to be accommodated, or longer life secured.

4.3.7 Snag free

Applications such as screw feeders handling fibrous products require a finish smooth to the touch so as not to snag on the product. Surfaces or components that are manually handled for routine dismantling or cleaned by hand, also require the contact parts to have a satisfactory 'feel' or texture. On surfaces against which loose solids are required to slide, minute snags or imperfections tend to present a key or foundation, on which a subsequent build-up takes place to cause excess residue or even a blockage. Details of grit size and surface flatness tend to be secondary to the treatment of welded junctions and the smoothness of exposed edges, except in the cases of a cosmetic finish.

4.3.8 Surface finishing methods

Many finishing techniques are available to meet these differing requirements. In some instances, a compound finishing process may be necessary to satisfy requirements that cannot be met by a common process. The most appropriate surface finish for a particular duty, and the method by which this is attained, are easier to determine and achieve when the objective for specification of a particular finish is clear. To prepare goods to a stated finish without knowing the reason for the selection of that finish, may well result in the expenditure of more time and cost than necessary, or an inferior standard than may be appropriate.

A range of surface finishing techniques is available to meet the aforementioned industrial needs. These are shown in Table 4.2.

4.3.9 Summary

The selection of a particular material of certain construction and finish may be made for a variety of reasons. Whereas material identification is verifiable with material certificates, unless the required finish is set out with some precision there are many opportunities for ambiguities and misinterpretations to arise. The attainment of some finishes is cost sensitive, so the interests of both manufacturers and users are served by the setting of a finish standard that is adequate, but not over-qualified for the duty in hand.

Table 4.2

Finish	*Description*
Mill finish	Softened and de-scaled (S & D), cold rolled (2 B), dull polished, bright polish, and mirror finish
Mechanical polish	Various grit finishes. Coarse to fine for differing degrees of surface smoothness
Blasting	Varied media, sand, glass, and plastic beads. An alternative to the construction from pre-polished sheets and mechanical polishing after fabrication is to bead-blast all stainless parts after fabrication, to bring the appearance to a common, pleasing standard. One drawback of bead-blasting is a tendency to 'fingermark'; hence some discretion is needed for its use on visible exposed surfaces that will be handled. Alternative blasting media are used to secure differing surface appearances and textures
Etching	Acid and wire brushing, normally for local treatment, e.g. welds
Electropolish	Surface 'de-plating' to expose virgin material. Another post-manufacturing technique is electro-polishing, where the entire component is placed in a de-plating bath and a surface layer of metal removed

One of the more difficult features to quantify in detail is welded joints and assemblies. A common way to indicate the standard of finish on welded stainless steel components is by the provision of small examples. It is comparatively easy to achieve any required finish on a small component that is fully accessible and manoeuvrable, but quite a different matter to attain the same quality over every minute region of a large and complex fabrication. To remove subjective judgement as far as practical from the inspector's task it may be useful to have examples of a slightly lower grade of finish, and/or of what is not acceptable at any point of the final equipment, for comparison.

A specification that identifies the basis on which a particular surface finish is selected will help the manufacturer to determine the appropriate construction, and provide some guidance on the method by which it is attained. Perhaps of equal importance, it will focus attention on details of the equipment design that are important to secure the objectives sought by the specification of a particular surface finish. The function of the designer is to match these attributes with other performance-related features of the equipment, and from these directions produce a blend of mechanical and geometric construction with a surface finish that meets the full requirements of the installation.

Chapter 5

Interfacing Screw Feeders with Hoppers

Screw feeders act as integrated components of the hopper serving to hold the stock of bulk material from which they feed. The hopper provides the volumetric holding capacity to allow the feeder to operate for a required period before replenishment. It also has the duty to supply a reliable feed of product at the rate demanded by the screw extractor, and in a condition which is suitable for the screw operation, the subsequent process requirements of the installation, and ultimate use of the product. Too often, the first function only is considered in any depth. Whether a bin discharge type of feeder or a metering screw feeder, sufficient hopper holding volume has to be accommodated to suit the application. Headroom is generally at a premium, if not a limiting factor due to site constraints. The first task of the designer is to establish the most suitable form of hopper. In order to make a balanced judgement on this it is necessary to consider the varied forms of flow regimes that prevail in hoppers, and build around the type encompassing features that need to be incorporated for the application in hand.

Screw feeders require a flood-feed supply of bulk material in order to control the rate of discharge. Apart from exceptional duties, such as extracting from a parallel supply chute, a bulk storage container provides the supply. A feature bearing on many aspects of materials stored in containers is the manner and order in which the differing regions of the contents fill and discharge. Various terms are used to describe the flow patterns by which these processes take place, the expressions 'mass flow' and 'funnel flow' or 'core flow', being widely used to describe patterns of flow behaviour. These descriptions relate to one important operating feature of the storage unit, but lack the precision to differentiate between other significant variations of behaviour in bulk storage systems. Flow patterns in hoppers have implications for flow prospects and for many other features of product behaviour and its condition.

The interface between the container outlet and the feeder inlet is a crucial region for storage facilities fitted with screw feeders, because the form of extraction pattern generated is inherent in the design of the feeder. This extraction pattern imposes an initial 'draw' profile of solids flow from the outlet region of the container. The relative rate of flow through differing cross-sectional regions of the outlet area dictates the manner in which the stored material can approach the outlet. This region is almost invariably a converging shape, set by a form of container construction that determines how the bulk material flows to satisfy the feeder demand. The size and

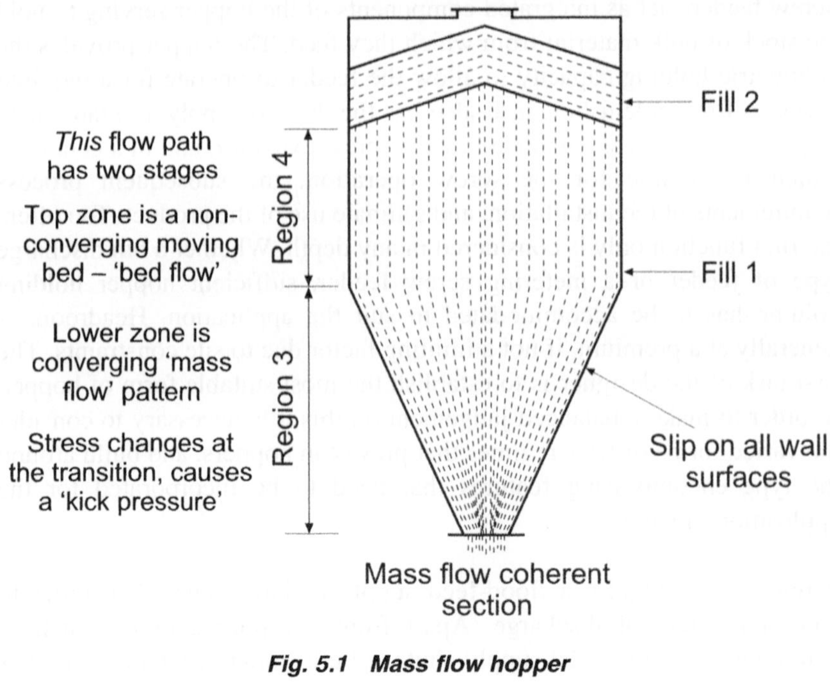

Fig. 5.1 Mass flow hopper

shape of the feeder inlet also has a great influence on the form and capacity of the container that it serves. For these reasons, the design of a screw feeder has an integral performance, and a structural and economic relationship with the total storage facility. By suitable choice of parameters, the designer can exploit characteristics of feeders to enhance and maximize key features of the installation.

5.1 Flow patterns in hoppers

The first and most important single feature of a flow pattern in a container is whether slip takes place on all contact surfaces between the contents and the container walls during a fully developed discharge condition. If it does, it is termed 'mass flow' by virtue of the movement of the entire mass (see Fig. 5.1). If it does not mass flow, it is often termed 'funnel flow' after the characteristic shape this type of flow channel takes in some cases (see Fig. 5.2), or 'core flow'. The first two definitions were laid down by Jenike in 1960, in his fundamental work on the gravity flow of bulk solids. Arnold Redler, in his UK and USA patents of 1920 relating to chain-type extractors, had previously defined the latter flow mode, and his term 'core flow' is common parlance in the UK. This form is sometimes referred to as 'internal flow'.

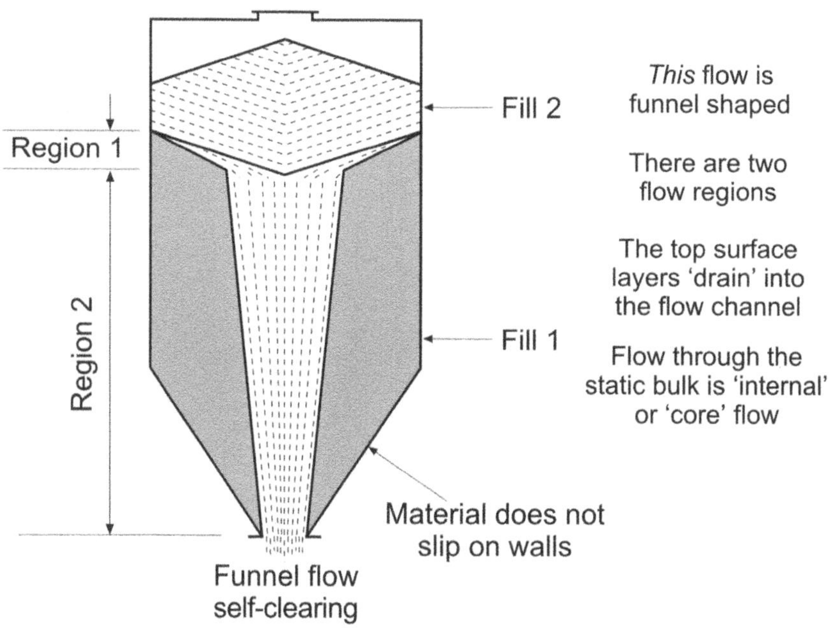

Fig. 5.2 Non-mass flow, self-clearing hopper (funnel flow)

However, within these expressions lie the seeds of simple misunderstandings. In some cases of what is called 'funnel flow', the flow channel looks nothing like a funnel. Its shape may vary from an unpredictable route drawing down from a flat surface in an obscure and unstable manner, as within a fluidized product, Fig. 5.3, to the other extreme where the flow channel has broadened to intersect the container walls underneath the surface of the stored contents, Fig. 5.4. The upper part, shown as region 4, moves in a bed flow manner, while the lower section, region 2, is distinctly core flow. This latter mode of behaviour may be externally assessed as a form of mass flow, but it most certainly is not.

The term 'core flow' is similarly misapplied to the overall flow pattern, whereas the 'core' section is usually only part of the material movement, as shown in region 2, Fig. 5.2. The top region of a 'core' flow channel is normally an unconfined boundary activity, shown as region 1, where material 'drains' on a repose surface into the confined flow section, but in some cases the upper section similarly expands to embrace the total cross-section of the stored product.

The term 'mass flow' as defined above is clear-cut, but it does not distinguish between converging channels, for which the expression was

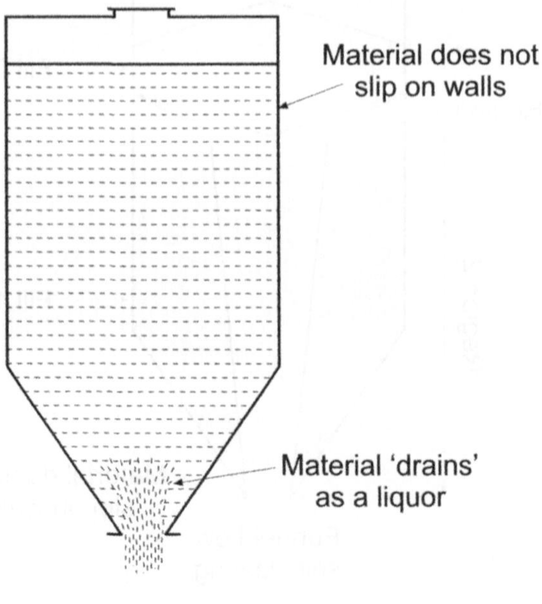

Fig. 5.3 Flow in fluidized hopper

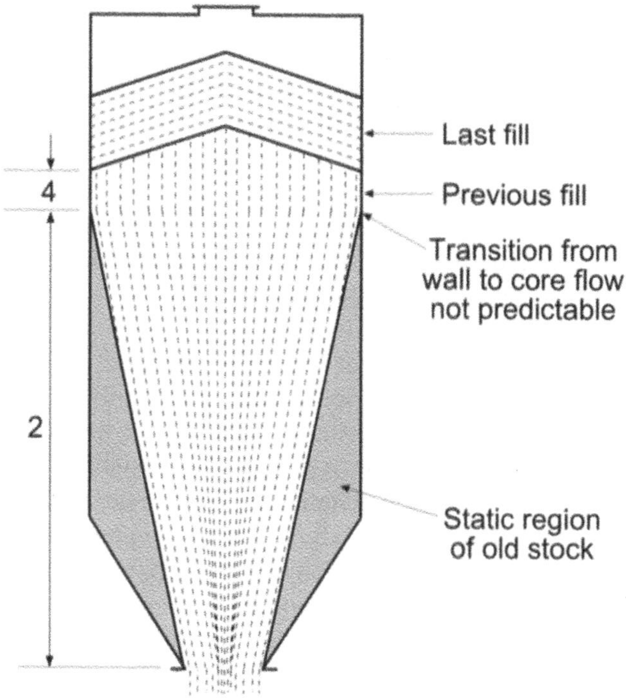

Fig. 5.4 Mixed flow pattern

originally coined, and the movement of material in a parallel channel as a surcharge above a converging section, shown respectively as regions 4 and 3 in Fig. 5.1. This is important because, within this flow pattern composite a dramatic change of stress condition occurs within the bulk material at the transition point. This feature has profound interest to the designers of bulk storage equipment and has been the source of many structural failures. The change of flow state that occurs in changing from a parallel to a converging channel, also takes place in other circumstances; therefore, there is value in drawing attention to the conditions in which the phenomenon occurs.

In practice, flow is either mass flow, or it is not. Describing the basic flow pattern as either 'mass flow' or 'non-mass flow' removes all grounds for confusion. It is proposed that these should be preferred expressions in the description of pattern of flow behaviour in storage containers.

5.1.1 'Mass flow' hoppers

A point to be emphasized is that 'mass flow' describes a flow pattern, and not containers of a particular geometry. Mass flow is a consequence of a particular combination of the flow of a specific bulk material, in a particular condition, that has certain frictional characteristics in relation to the geometry and construction of the flow channel in which it is moving. In some cases, the size of the outlet, or the rate at which material is allowed to flow from the container, will influence whether it discharges in a mass flow manner or not. The flow pattern that develops may even be sensitive to how soon the material empties after the container is filled, or to how deep or shallow the material is in the container. It is, however, a fundamental requirement of mass flow that the flow takes place over the entire cross-section of the container outlet.

The terms 'mass flow silo' or 'mass flow hopper' are only appropriate to specific application conditions, and only apply if the flow pattern generated satisfies the criterion of total slip of the contents on all wall contact surfaces during flow. A design that is established to generate mass flow requires that the bulk material, the contact surface of the material, and the operating circumstances, all remain within prescribed conditions for a given shape of container. If any of these parameters change, the flow pattern in the container may not develop mass flow form during its discharge, if the material discharges at all. Conversely, a container that normally discharges in a non-mass flow manner may begin to act in this way if the nature of the material stored, or other circumstances of operation change.

'Mass flow' is not necessarily flow with uniform velocity across the cross-section of the flow channel. Invariably, it is not in a converging flow channel, even though slip is taking place on the container walls. Thus, a commonly misquoted feature of mass flow as 'first in–first out' is an imprecise generalization. The implication that the sequence of fill is absolutely reflected in the sequence of discharge because the flow pattern is mass flow, is not matched in practice. In fact, the differential velocity across the section of a mass flow channel is exploited in some applications for the design of blending systems to mix stored contents.

The key feature of mass flow is that no regions of storage remain static during a cycle of discharge. This is guaranteed by its definition. The benefits which flow from this feature are the reasons that this form of flow design is often chosen by designers of storage systems, in spite of various disadvantages that may be associated with this form of flow.

5.1.2 General flow patterns in bulk storage containers

Definitions of flow patterns in bulk storage containers are set out in ISO/DIS 11697, as shown in Fig. 5.5. The modes shown in these diagrams

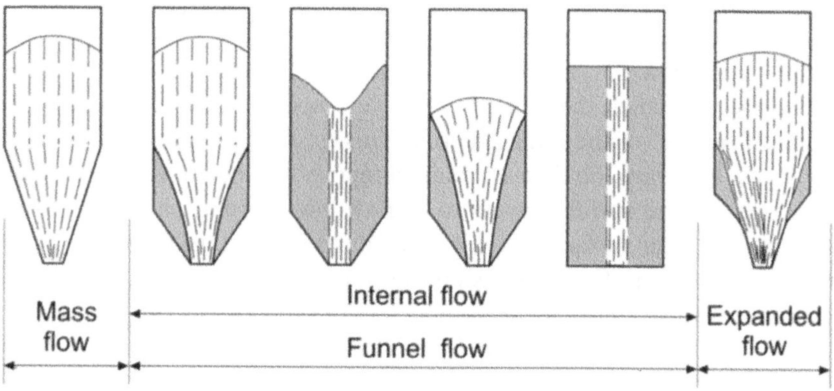

Fig. 5.5 Flow patterns as ISO/DIS 11697

do not clearly distinguish between differing combinations of regional zones of behaviour in a compound flow system, or unambiguously relate to the types of stress systems that are set up in the bulk material and on container walls. To describe the material behaviour more accurately, definitions of flow regime components are prescribed that allow a structured classification of differing flow patterns, characterized by the boundary conditions of the various flow channel forms.

The definitions adopted here embrace existing terms. They particularly accept the concept of a 'mass flow hopper' or a 'mass flow silo', as generic cases of bulk storage containers in which the whole contents move during the discharge process, with slip taking place between the product and all wall contact surfaces. As the term 'hopper' strictly relates to the converging section of a storage container, the words 'bin' and 'container' are henceforth used to represent complete bulk storage facilities, variously described as hoppers, silos, bunkers, bins, vessels, and storage devices for bulk materials. These apply to all shapes and scales of use, and the varied means for controlling filling and emptying.

As distinct from descriptions of global flow regimes, which are composite patterns applying to the whole container, there are four basic ways in which bulk materials move as regional flow channels, or 'zones of behaviour', as a bulk storage bin empties: repose flow; core flow; converging mass flow; and bed flow. These are described and defined in the following sub-sections. Some of the figures quoted follow later in the text, under descriptions of global flow regimes.

Repose flow
This is essentially an unconfined surface behaviour, where layers of material slide down over an inclined static bed of product. The slope of the bed surface at which the material will slide upon itself is defined as the 'angle of repose' of the bulk material. This is shown in the top region of Fig 5.2. In this form the flow is 'drained repose', as the bin is emptying by sloughing off the surface layers to a drawdown through a static bed of product. A similar surface effect is given during filling from a single point entry stream, where the contents build up in a conical pile as the material cascades down the growing surface as 'poured repose'.

Material entering a storage vessel as a steady stream does not necessarily spread out to an even film over the cone of fill, but more usually finds a radial region of reduced resistance down which a surge will occur as a local avalanche. The deposition of a fresh layer of material then places an obstruction against continuing flow in this direction and the in-coming stream finds another radial outlet in which it can flow as a new avalanche. The erratic surging of material in differing radial positions progressively forms a growing surface at a consistent angle of slope. The main features of interest lie in the degree of ullage to be allowed for loss of volume above the pile around the point of fill, and the positioning of level indicators on the walls to reflect the hopper contents.

A bulk material which behaves in a fluid manner when very dilated, will flush out to a level surface of fill, as can be seen in Fig. 5.3, and display hydrostatic pressure conditions until settled. Poor flow materials, such as damp and cohesive powders, achieve steep filling repose angles. Therefore, appropriate allowances must be made in calculating holding volumes and fixing the location of level indicators. Coarse, firm, non-adhesive products tend to settle quickly, to form a stable bulk structure, when they come to rest after pouring. Such bulk materials attain consistent density conditions and repeatable 'angles of repose', virtually independent of the manner in which the slope of the pile is formed.

Bulk materials that are composed of fine particulate constituents are much more sensitive to the influence of the void gas, as changes of volume are resisted by the rate at which the gas can progress through the interstices of the mass. As a result, the repose conditions are very dependent upon the conditions of slope formation, as is the density condition to which they settle. Such materials do not have a specific 'angle of repose', either in filling or discharge conditions, and it is misleading to ascribe a definitive value to them without specifying very closely the circumstances under which the slope is formed. In general, it is better to advise that such materials do not have a consistent repose condition. When a fine, dry product is filling into a container the surface may vary from a relatively shallow angle to zero.

Repose flow taking place during the emptying of a bin is usually at a different surface gradient to the inclination during filling. With a bulk material that gains strength by compaction, the manner in which the surface collapses can vary widely. Such materials will tend to fail in steps, as steep initial repose conditions become unstable as the height of the 'wall' increases. The collapsing bulk may then develop a rough form of dynamic repose with a surface bounded by weak agglomerates. In extreme cases, the strength is sufficient to hold a vertical cliff. The central core of flow in a non-mass flow hopper then empties as a hole through the bulk, to leave a stable 'pipe' or 'rathole' and consequent flow stoppage. The fill profile of the stored surface is not of key design interest for flow. It is, however, relevant to the bin contents and for level indication, particularly where detectors are positioned against the side walls of the bin. For a given level of material against the wall, variations of surface profile that range from a deposited cone to a steep drained cone represent large differences in capacity, see Fig. 5.7.

The length of the filling slope has a major bearing on the degree of segregation caused during the filling process. Various separation mechanisms are active in an inclined stream of flow to favour the deposition of fines and the continuation of the coarse fractions. The net result is that fines normally tend to collect in the centre of the pile, and larger granules and lumps congregate around the periphery against the walls. In some circumstances, however, the converse applies. The ultimate consequences of any uneven distribution of the constituents then depend heavily upon the pattern in which the stored contents discharge. A common misconception is that a mass flow hopper will re-mix a material that has segregated during the filling process. This process will generally apply to the initial and main portion of the stored contents. A consequence of the velocity contours which develop in a converging mass flow section is that the terminal portion of contents discharged from a mass flow container are drawn from the upper, outer annulus of the stored material. This is a region in which the effects of repose-type segregation are most pronounced. Hence towards the end of the discharge process, the material will tend to contain a higher than average proportion of the coarse fractions.

Core flow (sometimes referred to as 'internal flow' or 'pipe flow')
This is a flow channel that is bounded by a static region of stored material, as shown in the lower regions of the stored material in Figs 5.2 and 5.4 and the middle regions of Fig. 5.7, 5.8, and 5.9. The flow channel is usually of a converging form, a reason why the term 'pipe flow' may be misleading or lead to ambiguity.

This form develops when the walls adjacent to the outlet are insufficiently steep for confined bulk material to slip on the wall surface. The flow channel tends to diverge slowly, from the outlet, according to the properties and condition of the bulk material. Unless the material surrounding the flow channel holds together to form a stable 'pipe' or 'rathole', the local surface depression allows adjoining material to collapse to form a drained cone of repose. The 'core' channel passes loose material from the surface layers, causing the drained cone size to increase and, eventually, to reach the walls. At this stage, the level of material against the walls reduces, with the surface layer of the cone sliding over the static bed into the core channel. The characteristic shape of the combined repose layer and core flow led Jenike to term this overall pattern of behaviour 'funnel flow', see Fig. 5.3.

Converging mass flow
This form of flow occurs when the cross-section of the flow channel extends to the confining walls of a container that has a converging cross-section. Slip takes place on all wall contact surfaces during flow, and the material in the flow channel is in a state of total deformation. This type of flow takes place in a hopper as shown in Fig. 5.6 and in the lower regions of the containers shown in Figs 5.1, 5.7, and 5.8.

The term 'mass flow' was coined to describe converging flow with slip taking place on the boundary surfaces of the confining container. The meaning of the term has been extended in use, to describe the situation of total movement of bin contents. This essentially requires the material to slip on all contact surfaces of the container, whether converging or not. In all situations the outlet of a bin must be active over the whole cross-section to enable mass flow.

To avoid confusion with the non-converging movement of a body of material in a parallel flow channel, the term 'mass flow', when used to describe a flow channel, should only apply to flow in a converging channel with slip on the boundary walls, as shown in Fig. 5.7. To apply the

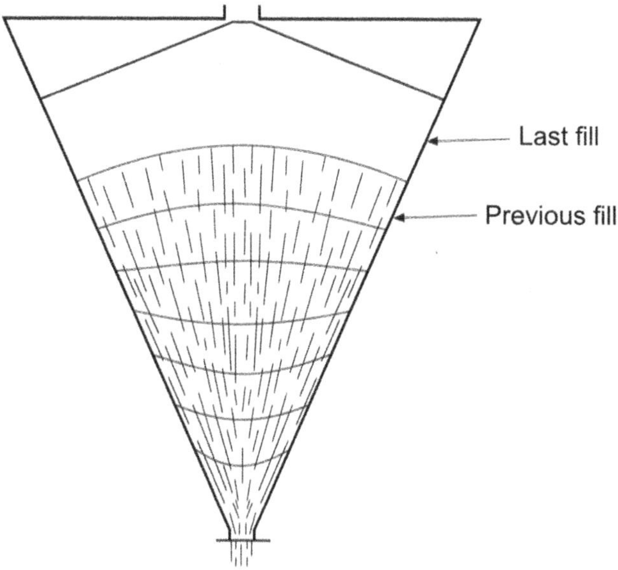

Fig. 5.6 Converging mass flow

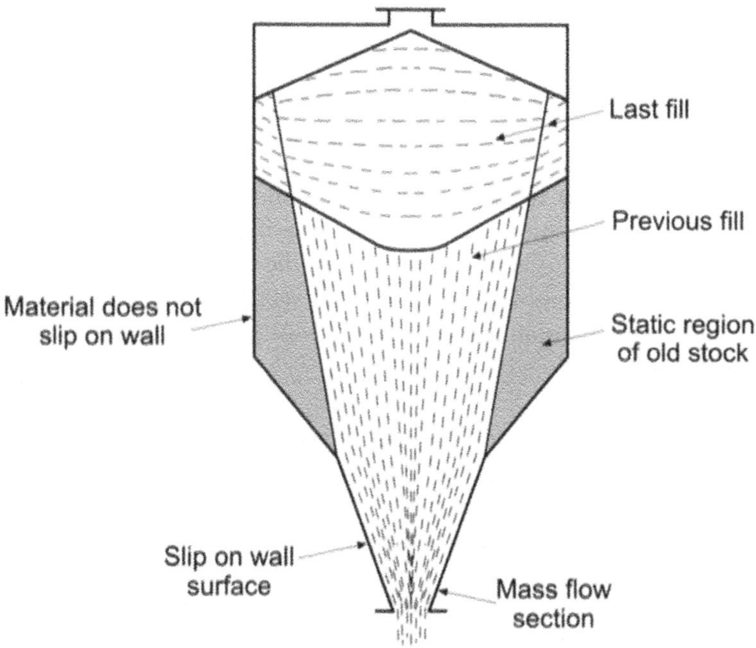

Fig. 5.7 Expanded flow

expression 'mass flow hopper' to a container constructed with both converging and parallel storage sections is strictly a contradiction in terms, as both 'hopper' and the only defined meaning of 'mass flow' relate exclusively to a converging vessel. However, by common usage the terms 'mass flow hopper', 'mass flow silo', 'mass flow bin', and even 'mass flow pattern', apply to a complete storage unit in which the contents slip on all wall contact surfaces during flow. The need to describe the situations where a portion of the flow channel slips on the walls of a parallel channel is met by the term 'bed flow'. The absence of change in cross-section allows motion of the bulk without internal movement, i.e. the material can move as a coherent mass in transit, or it can have different flow velocities across its cross-section.

Bed flow

This occurs where the cross-section of the flow channel extends to parallel confining walls of the container, i.e. the flow channel is neither converging nor diverging, and slip is taking place on all the contact surfaces during flow. The flow velocity across the bed will tend to increase in the central regions as it approaches the transition to a converging section, whether this be a converging mass flow or a converging core flow channel. Bed flow is shown in the top regions of Figs 5.1, 5.4, 5.8, and 5.9.

When a body of material moves en bloc it is not flowing in the conventional sense of the word. However, the term 'bed flow' is used here to describe material moving over the whole cross-section of a parallel flow channel while sliding on the boundary container walls, and complements the descriptions of flow regimes for the common types of hopper flow patterns. The cross-section does not necessarily move with uniform velocity; this depends to a large extent on the underlying drawdown pattern at the termination of the parallel wall section, as modified in the parallel channel by the relationship between wall friction and the internal friction of the bulk.

The afore-mentioned types of flow take place in various combinations, according to the geometry of the bin, and the properties of the bulk material, to give various global patterns of behaviour. Common forms are shown in the various figures. Complications arise when the flow channel is offset from the centreline of the container, or flow becomes eccentric to the container walls. Flow zones and channels follow similar lines to those of concentric systems, but stress evaluation in these cases is usually the domain of the expert.

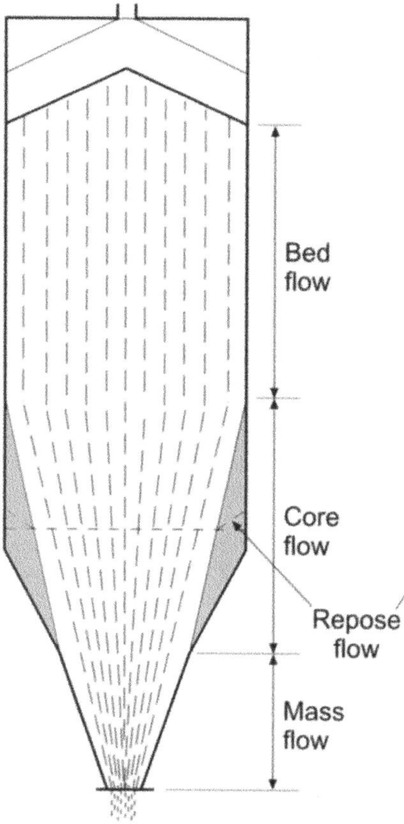

Fig. 5.8 Expanded, mixed flow

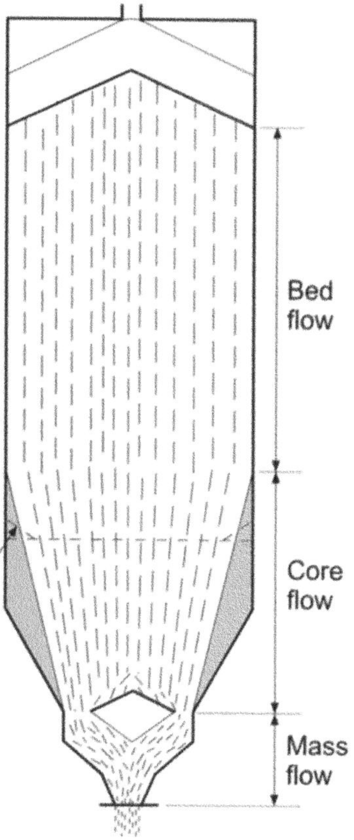

Fig. 5.9 Pre-expanded flow

The different repose conditions between emptying and filling has a major influence on the location of fresh contents in a partially emptied hopper, fig. 5.10. Combined with the nature of the extraction pattern this has significant implications as to the maximum residence time that some of the contents experience in different sequences of discharge and refill.

5.1.3 Patterns of outflow

The order in which different zones of the stored material empty affects the residence time of the material filled into the various regions of the container contents. When discharge commences, it invariably takes a little time for a complete pattern of flow to develop through the body of material. Initially a wave of 'dilation' is propagated from the outlet through the settled bed until, in a hopper of reliable flow design, a steady state of flow has been established. When discharge stops, the dilated flow channel settles progressively to a new settled condition.

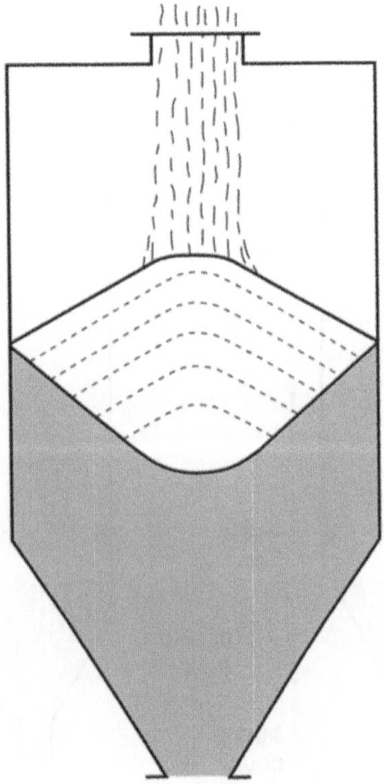

Fig. 5.10 Repose fill contours

The manner in which core flow develops is that material falling from the outlet is replenished by material falling from the static bed. It is much easier for overlaying material to fall into the vacated space rather than to move sideways from the mass. Consequently, the flow channel spreads only gradually in cross-section, if at all, until it breaks through to the upper surface of the contents. The way in which the surface layers drain into the flow channel is then determined by how readily it will deform against the unconfined surface of the exposed 'core'. The drained repose angle of the material is a measure of its settled strength. The characteristic 'funnel' shape of the flow boundary is thus developed.

The 'funnel' flow channel essentially consists of two component parts: (i) the drained repose or unconfined surface layer, and (ii) the 'core flow' channel confined by the static bed. Should a hopper with this pattern of discharge be refilled before it has emptied, some of the original material

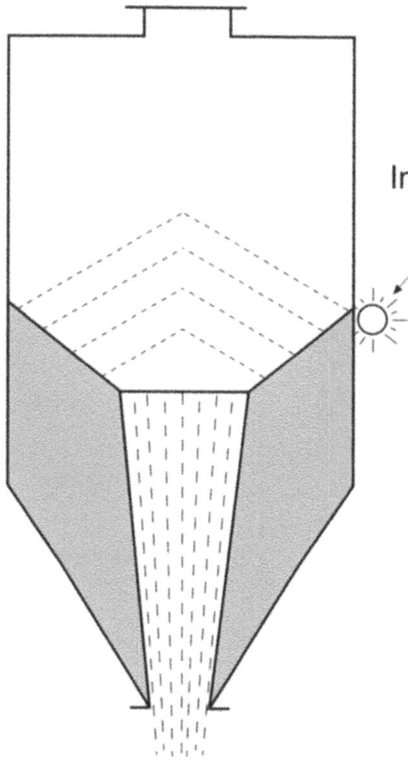

Fig. 5.11 Inconsistent level indication

will stay in the hopper until the hopper has totally discharged. Thus a number of refills will restrain the containment of a portion of the first fill. Depending upon the level at each time of refill, there is much uncertainty as to what portion of what filling batch is being discharged at any time. There are also serious implications with materials that segregate, as the boundary surface of refill forms a layer that will discharge in a concentrated form when the level falls back to that at which the refill commenced.

The surface contour difference between filling and drained repose can also reflect considerable differences in the capacity stored, Fig. 5.11. A design process for storage hoppers is shown in Fig. 5.12. The various advantages and drawbacks of both mass flow and non-mass flow are shown in Table 5.1. Charts indicating the wall conditions that determine which form of flow will prevail in given circumstances are given in Figs 5.13 and 5.14.

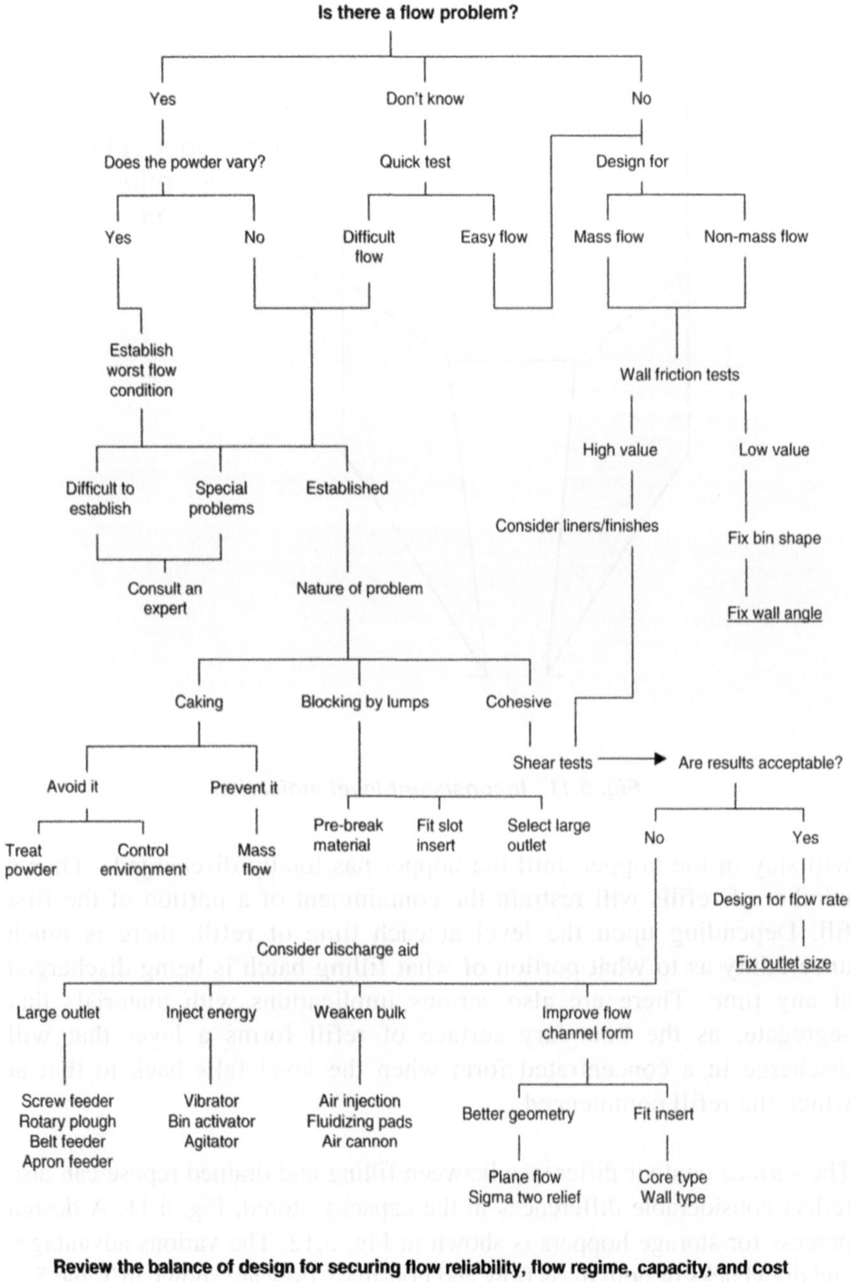

Fig. 5.12 Hopper design sequence

Table 5.1 Advantages and drawbacks of mass flow

Advantages	*Drawbacks*
No 'dead' regions of flow	Tall headroom/reduced storage capacity
More predictable storage times	Potential wear on walls
Secures flow through smaller outlets	The outlet must be fully 'live'
Generally reduces segregation	Powder tests are essential
Resists 'through-flow flushing'	High wall pressures are generated at the hip joint
The flow pattern is predictable	The design relates only to the condition of the product tested
Flow can be exploited to blend contents	Any property change may negate mass flow
Proven design guarantees reliable flow	Flow rate is less than non-mass flow with same outlet

Notes The flow velocity is *not* uniform across the converging section of the hopper. Mass flow at the outlet region only (mixed flow), secures orifice size benefits

5.1.4 Global flow regimes

Mass flow This is where all the contents move during discharge. It follows that wall slip takes place on all boundary contact surfaces as shown previously in Figs 5.1 and 5.6.

Funnel flow This term describes combined system of core flow and drained repose, where no portion of the flow channel is slipping on the container walls, see Fig. 5.2 shown earlier.

Expanded flow This is a flow pattern comprising of a section of mass flow behaviour adjacent to the outlet, on which is superimposed a core flow section, as shown previously in Fig. 5.8. The upper region has the characteristics of funnel flow behaviour. This pattern should not be confused with a flow channel pre-expanded by use of a discharge device, such as a bin activator, as shown previously in Fig. 5.9, or a multi-screw feeder. These are normally fitted to secure a large outlet opening to avoid arching or control the discharge rate, rather than achieve a form of flow channel.

Mixed flow This term describes a situation where a core flow channel diverges in the stored contents to meet the boundary walls of the container at some distance under the surface level of the contents.

Expanded, mixed flow This is a combined pattern of a lower mass flow region, with a core flow mid-section that expands to the container wall as a bed flow upper region.

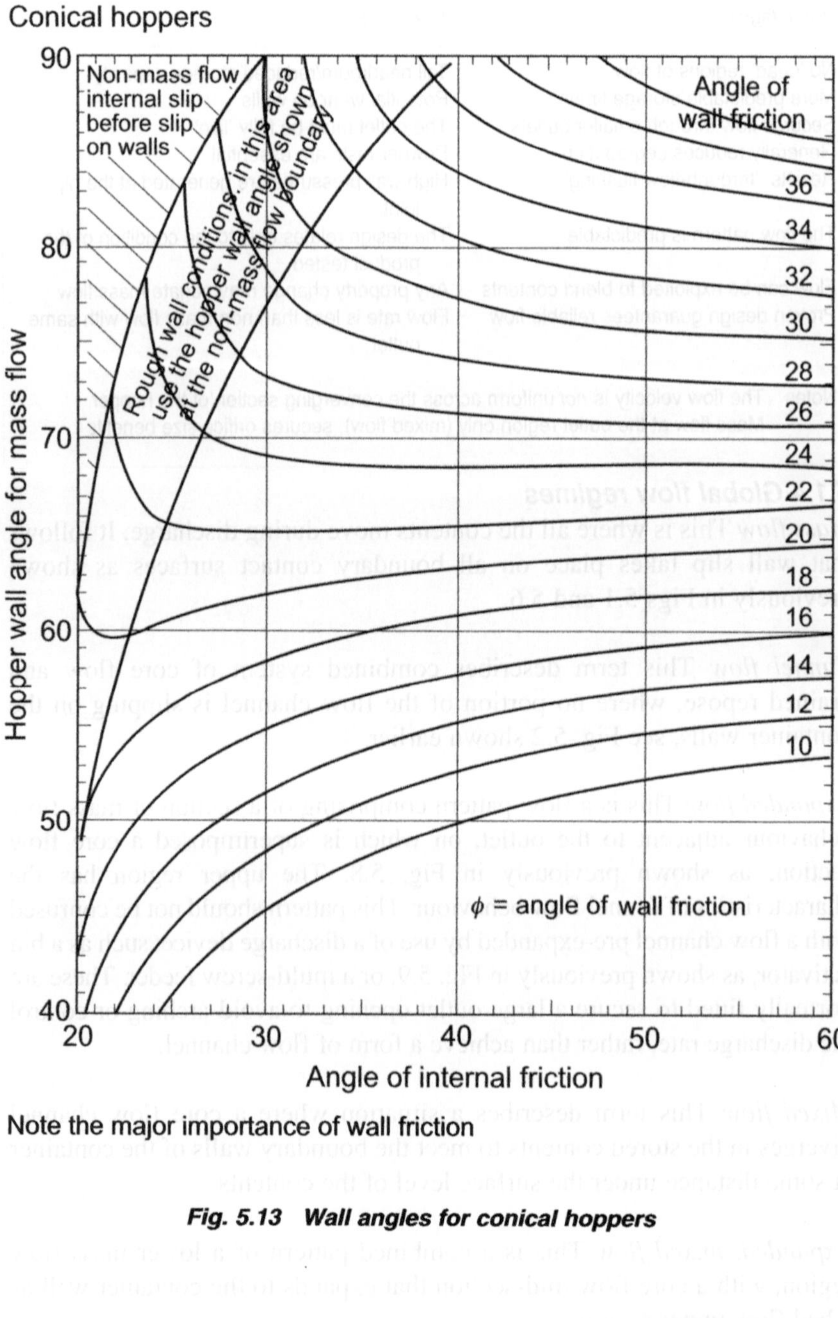

Fig. 5.13 Wall angles for conical hoppers

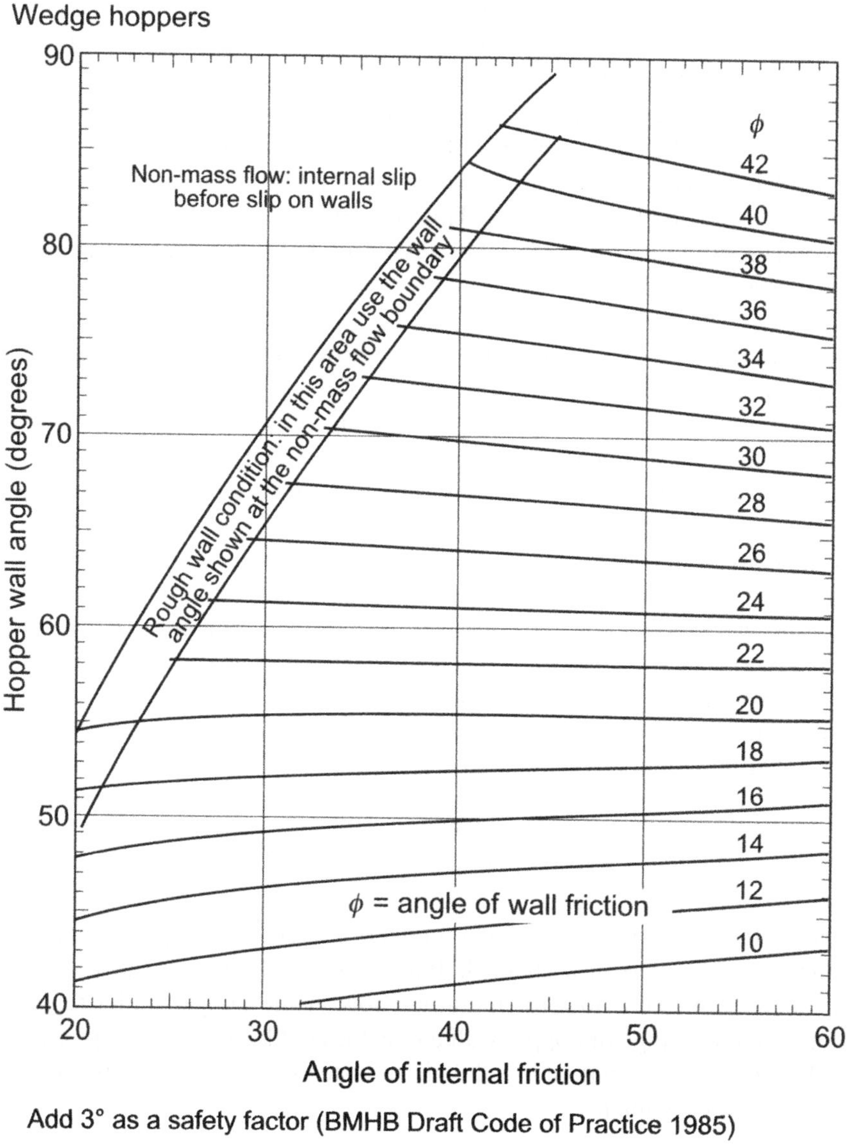

Fig. 5.14 Wall angles for 'vee' hoppers

Eccentric flow This term describes the flow form when the cross-section of the moving stream is partially bounded by a static region of bulk product and the remaining periphery is moving against the inner wall of the container (not illustrated). This class of flow is usually connected with features of eccentricity of the system and requires expert evaluation as to its effects. An important distinction arises between two patterns, both of which involve the entire contents of the container moving during a period of discharge. The simpler case is where the whole contents are converging uniformly in mass flow, Fig. 5.6. This comprises a smooth process of bulk deformation and both the internal stresses acting on the material and the wall pressures vary smoothly.

By contrast, in a container such as that shown in Fig. 5.1, the material changes from a bed flow region, as the walls of the flow channel are parallel, to a mass flow region of converging flow. For this flow pattern there is a radical change in the stresses acting on the material at the transition between the two regimes. The top part is pressing against the walls with an 'active' pressure, reflecting the filling conditions. At the start of the converging section of the hopper the material is required to deform internally to reduce its cross-section. The forces required to provide internal shear of the bulk and commence this deformation process generate a 'passive' state of stress in the material resisting deformation. The change of stress condition is dramatic and referred to as 'kick pressure' on the walls. In large silos, and particularly those of concrete construction, special structural consideration has to be given to these loads.

A construction utilizing a mass flow region in the lower part of the hopper and a core flow construction above is termed an 'expanded flow' construction. It is so termed because the material will slip on a smooth wall surface at a lower angle than on itself in a core flow type of hopper, hence expanding the flow channel. It may also be employed to provide the advantages of mass flow for initiating flow at the hopper outlet region. Note that the change of hopper wall angle does not lead to a 'kick' pressure at this point, as the material is already in a steady state of convergence in the core flow channel.

When any core flow channel spreads to meet the walls of the container below the surface level of the stored material, the upper region of the flow channel is of bed flow type. This combination is termed 'mixed flow'. This is an important form to recognize, because a 'kick' pressure will be created at the transition where bed flow changes to a converging core flow channel, see Fig. 5.4. The prospect of creating this type of transition is greater when the size of the effective outlet of a container is enlarged by means of a feeder or a bin activator, because the flow channel does not have far to spread to meet the container wall. A particular danger arising from this change of flow stress condition involving a core flow channel, is that the location of the intersection between the flow channel and the wall cannot be reliably predicted, and it is not necessarily stable. In the uncertain conditions of a 'floating kick pressure' virtually the whole depth of the container wall must be designed to allow for extreme variations of the high load location.

While most of the terms recommended are widely accepted, common misconceptions persist, particularly with regard to 'funnel flow' and the different zones of behaviour in a 'mass flow hopper' that incorporates a section of non-converging boundary. General use of the above definitions of flow regimes and their local zones of composite flow behaviour will lead to a better understanding of the various mechanisms that are active during flow processes, and also secure a common international basis for descriptions in technical publications.

5.2 Screw geometry

There are various techniques used to increase the extraction capacity along the axis of a screw. The most influential is to increase the screw diameter, as the cross-sectional area increases as the square of this dimension. Other ways are to increase the pitch, reduce the centre shaft diameter, or reduce the friction on the face of the screw flight. These variations can be made in virtually any combination, or independently. The effect of these changes will be examined in detail.

5.2.1 Screw diameter change

Although effective in varying the extraction this has two main drawbacks: the most significant is that the width of opening served by the screw changes from the small diameter end to the larger diameter. The second drawback is that, unless a wide gap of static residue is to be left at the side of the small diameter region, the casing must follow the conical form of the screw, which incurs an expensive form of fabrication. The wall of the screw casing must align with that of the hopper wall to avoid boundary discontinuities such as 'steps' or ledges. An undesirable feature of taper screws is that the 'small' diameter end must be sufficiently wide to avoid the formation of a stable arch over the slot. It is also essential for the full width of the outlet to generate a 'live' flow pattern.

It is not common practice to design a hopper outlet slot with a taper form, so the normal method of accommodating a taper screw, is to flare the feeder casing at the small diameter end, to a width similar to the casing at the larger diameter end. Almost invariably the wall angle at the small end of the casing is not adequate to provide slip. This is partly due to the inclination usually being less than the required mass flow angle, and partly because the static residue resting in the lower clearance space between the screw flight and the casing offers an obstruction to slip down the casing side wall. The smaller diameter of screw at one end of the hopper outlet must also be larger than the critical arching span of the material. If there are side regions of static product due to this flow impediment of the casing, the resulting effect is to negate the prospects of wall slip on the hopper outlet walls, and hence prevent mass flow, see Fig. 5.15.

To overcome these problems great care is needed in the design and construction of the feeder casing. One way would be to reduce the interface width of the hopper to suit the smallest diameter of the screw, but this is rarely undertaken. A more efficient method is to form steep side walls, such that the material develops an inclined shear plane from the residue layer under the screw to the casing wall, below or at the interface to the hopper outlet, Fig. 5.16. When matching feeders to non-mass flow hoppers this problem does not arise, although the opening width must still exceed the arching potential of the material and, if there is no wall slip, this size must be based on non-mass flow design.

5.2.2 Pitch change

No casing shape problems arise when alterations are confined to the screw pitch, so a common and economical manufacturing technique is to make variations of pitch only. Pitch changes can be made at every flight spacing, or at intervals along the screw length. To serve mass flow hoppers the changes should be continuous. There are, however, limits to the changes that can be made. If the pitch is very short, there is a danger of material clogging in between the flights due to the deep and narrow gap formed. At longer pitches the efficiency of transfer falls off and face pressures become excessive because of the length of the plug of material that has to be pushed forward. As a result of these limitations, only a limited range of pitch variation is practical. The extraction pattern is also not uniform along the length of a graduated pitch screw. At the 'start' of the screw, the screw volume fills and, because subsequent increments tend to be quite minor, only small, marginal increments of volume are available for further inflow.

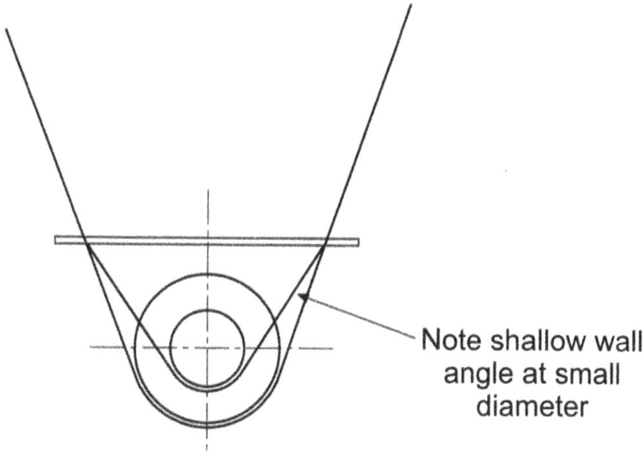

Fig. 5.15 'Flared' casing for taper screw

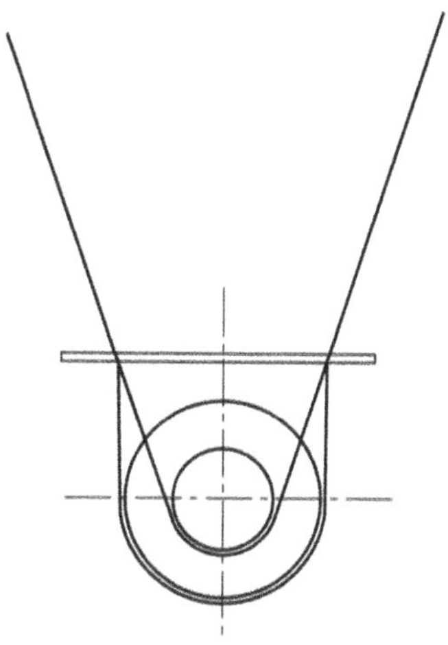

Fig. 5.16 Steep casing for taper screw

The construction of the choke section at the 'exit' end of the screw interface region can allow a degree of extra material to 'escape' with the screw. This generally leads to large variations in flow velocities across the length of the hopper outlet, but as the two regions of greater extraction are against the end walls of the hopper, the flow channel is generally effective.

For mass flow hopper applications, these features limit the effective length of screws that vary in pitch only, to about five or six screw diameters. For non-mass flow applications, pitch changes are a useful means to reduce power and secure an improved extraction pattern, and much longer exposed sections of screws can be used. These benefits may not be essential, but offer advantages by avoiding excessive 'dead' zones of storage.

5.2.3 Shaft diameter change

Reducing the diameter of the centre shaft of the screw increases the effective area for material transfer. The effect is far from proportional to the area of the shaft cross-section because the flight form in the central region of the screw is constructed at steeper helix angles than at the periphery. However, in combination with pitch changes, the geometry allows greater variation without the risk of 'logging' due to deep, narrow gaps between the flight pitches. When the diameter of the centre shaft is large in relation to the outside diameter of the screw, it is good practice to adopt short pitch construction. Long, shallow pitches are not efficient to promote the motion of narrow columns of material. Balanced variations of centre tube and pitch sizes can give effective transfer progression, and hence extended live interface flow from slot outlets, Fig. 5.17.

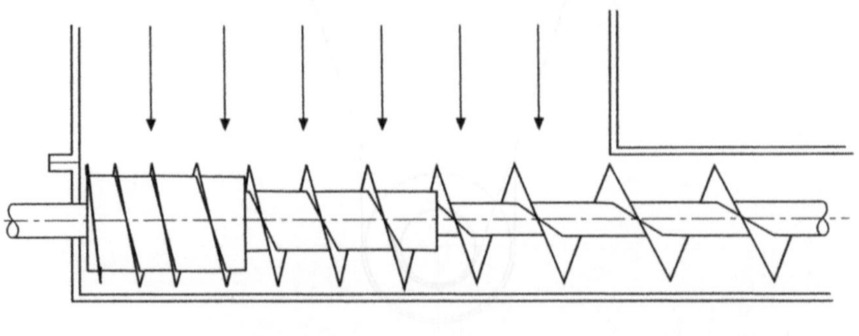

Fig. 5.17 Combination of shaft diameter and pitch change

5.2.4 Ribbon and 'shaftless' screws

Although normally specified to deal with wet, sticky, and cohesive products, ribbon screws (as shown in Figs 5.18 a and b) can be effectively used, in whole or in part, in feeder construction. A small gain in capacity is given by this change of design, so a final section of ribbons on a variable pitch feed screw, enhances the extraction rate to make the flow channel more effective. Screws without a centre shaft are made for relatively crude

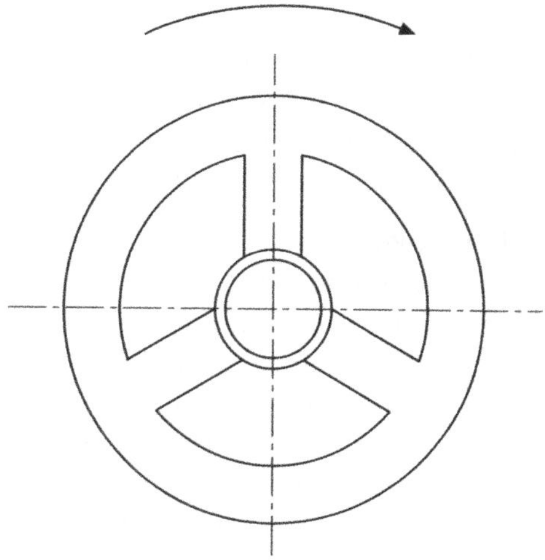

Fig. 5.18a Conventional ribbon screws

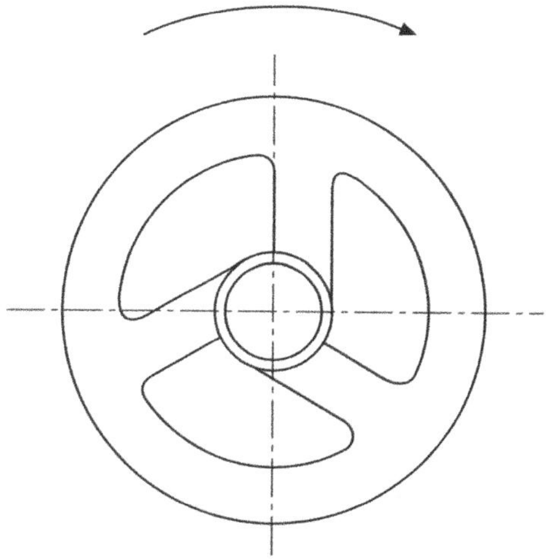

Fig. 5.18b 'Ajax'-type ribbon screws

conveying duties. Their ability to resist clogging outweighs the lack of bearings (they run on liner plates or on the casing of the conveyor).

For feeding applications, shaftless screws are generally short and cantilevered from a drive shaft, which in turn may incorporate a short section of ordinary screw. Very small diameter screws, 10–30 mm in diameter, are often run in tubes without regard for casing contact because their weight is so small. Larger screw diameter sections are usually of short length and part of a cantilevered arrangement, utilizing the shaftless part as a final increment of extraction in the hopper interface. Such screws are used singly, or in multiple screw feeders Fig. 5.19.

5.2.5 Paddle-blade screws

The most common use of screws made up from segmental blades is to incorporate a degree of mixing or pre-breaking into the machine. Blades can be placed in a helical pattern to form a crude screw, Fig. 5.20a, or arranged in a counter helical pattern to give increased mixing as in Fig. 5.20b. The segments can be welded on to the shaft or mounted on adjustable rods fitted through the centre shaft, to allow the angle of the paddles to be set individually. Bolted-on members are replaceable in the event of excess wear or damage. They are also relatively easy to construct, which simplifies field maintenance. In general, loose-blade machines of this type are crude for straightforward feeding duties, but are more usually designed with extra functions in mind.

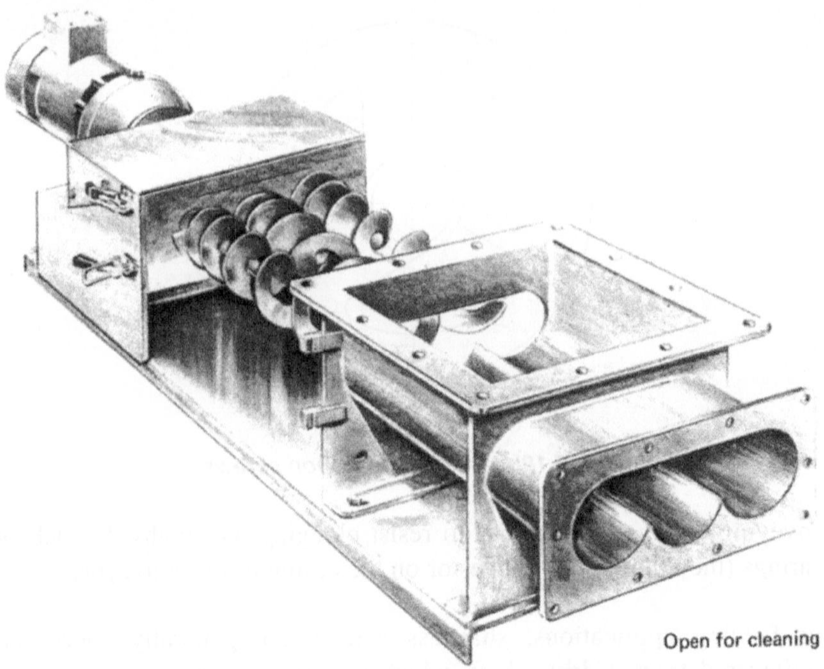

Fig. 5.19 Multiple, cantilever feed screws, with shaftless sections

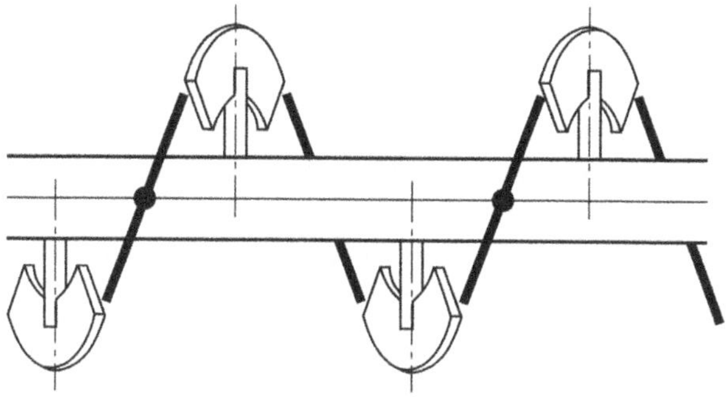

a Helical mounted paddles

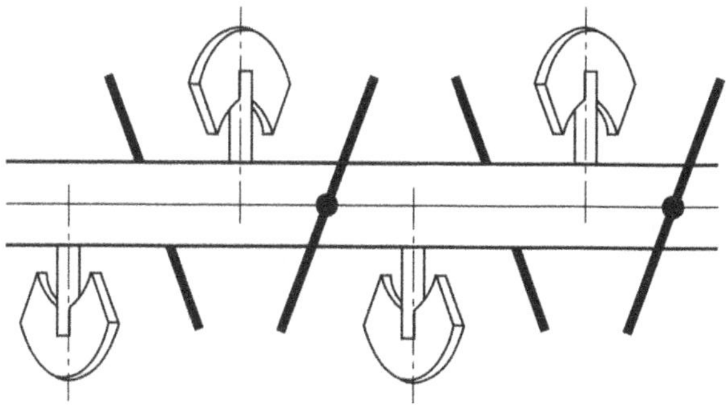

b Counter-helical mounted paddles

Fig. 5.20 Helix and counter-helix mounting of paddle blades

5.3 Feed hopper geometry

The design of a storage container for bulk materials is rarely taken in isolation, normally being influenced by various facets of the associated plant or the background of the manufacturer. There are three main steps to selection of storage hopper geometry: (i) body configuration; (ii) outlet size and shape; and (iii) the transformation between the two.

The first decision is whether the body of the hopper is to be round or rectangular. In terms of capacity there is little difference if the final outlet is small and symmetrical. The gain in cross-section of a square or rectangular bin over one of circular construction, is largely negated by the extra height required for the transformation section to suit any given wall slope, whether this angle is for self-clearing or to suit a mass flow design.

In the case of a non-circular section, the inclination of the gully angle determines the height of the transformation between the body and the outlet. Figure 5.21 illustrates the relative storage volumes of different container cross-sections. The virtues of a rectangular shape only start to show large benefits when the geometry of a long slot outlet can be exploited to offer increased capacity. This is because lower wall angles can be used for plane, rather than radial flow, and the gain in storage volume is proportional to the length of the slot.

For large storage silos, a circular construction offers various structural advantages, as it is better able to sustain internal pressures than flat walls. This is, therefore, a common form for individual bins of large capacity. Very tall, nested silo structures, for light duties such as flour storage, using aluminium construction, are often of rectangular design using slot outlets with screw feeders to secure maximum holding capacity, and mass flow of the contents within a given cross-section.

The second, and crucial decision, is the shape of the outlet. This is determined by the feeder connection, of a size rated to deliver the output, with an extraction profile and inlet length as selected. The interface connection for a screw feeder is usually a slot form, with a length

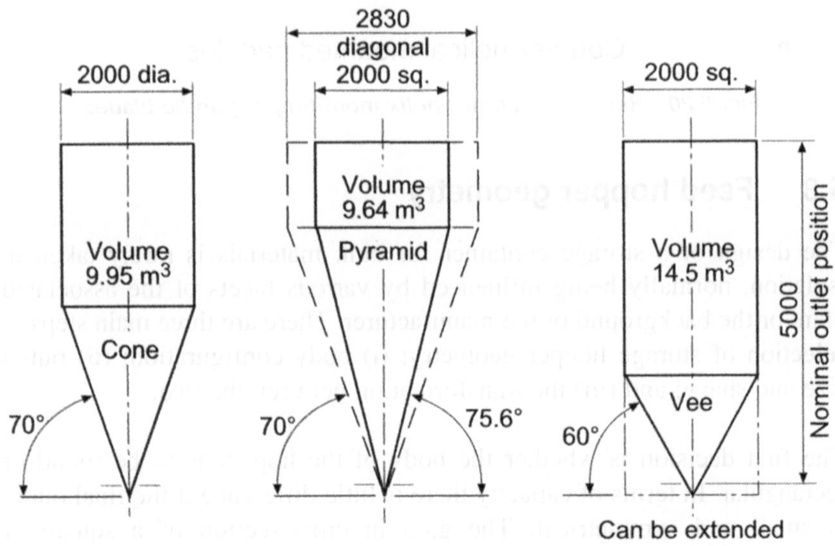

Fig. 5.21 Relative holding capacity of conical, pyramid, and 'vee' base storage bins

exceeding twice the width of the outlet. Shorter inlets are occasionally used, but these neglect to take advantage of the ability of screw feeders to serve slot outlets that offer better flow prospects and greater storage capacity. The fixing of one key parameter tends to influence the rest. For example, deciding an outlet size for flow considerations requires a width of feeder to serve the opening size selected. It is usually obvious whether a single screw can meet this task, and what speed it would have to run at to deliver the rated discharge. The section of a slot length to acquire a given storage volume without shallow end walls, puts a limit to the lowest size of centre tube that can span between the end bearings. The tube size then sets a lower boundary to the size of screw that may be used, without creating a screw form out of normal proportions.

The third important step is to fix the form of the transition between the body section and the screw feeder inlet. The transition is relatively simple to determine for non-mass flow duties. The only operational feature is that the material must self-clear from the walls and any corner gullies. Square corners tend to hold a small radius of any but the most free-flowing of solids. To avoid residue, it is good practice to radius gully corners. The potential for an arch to form is much increased when static product around an outlet provides a rough-faced base for supporting the arch foundations. The width of opening to guarantee reliable flow through a static bed of material has to be sized on the critical span dimension of a 'rathole'. Further, a flow channel that forms within the contents of the bin is relatively narrow; therefore, only a small proportion of the total mass is in motion and the surface layers are progressively extracted. There are advantages to be gained by the use of a progressive extraction profile with a non-mass flow bin. Apart from a reduction in drive power and more consistent 'fill' conditions of the screw, when the flow channel has a 'slot' shape the prospect of a 'rathole' forming is low, because flat sides of a flow channel are less stable than a circular hole.

For a material that is not free-flowing, a short mass flow section should be used leading to the outlet opening, as this provides flow through the smallest practical size or outlet. A mass flow slot outlet, with a slot length more than three times its width, is as effective for flow as a circle with a diameter twice the width of the slot. Flow will also take place through a mass flow opening at half the size of a non-mass flow opening. As a result, a mass flow slot is as effective for reliable flow as a circular non-mass flow outlet four times its width. An expanded flow construction for this purpose need only be extended high enough for the width of opening to avoid arching in non-mass flow conditions, i.e. above the size of a stable 'rathole', which for a slot is twice the width of the mass flow slot outlet.

Note that the advantages of selecting a wall inclination for mass flow on the basis of a 'vee' hopper can only be achieved if the ends of the slot are vertical. If the end walls are converging then the wall inclination has to be based upon the gully angle as a cone, which is normally a very severe penalty. Above the mass flow section, the wall inclination merely needs to provide a self-clearing angle. The contrast between mass flow and self-clearing angles may appear excessive. Intuitively, inclinations of 60 and 45 degrees may seem realistic, but 70 and 35 degrees are probably more appropriate in most cases. The actual angle specified should be calculated according to wall friction tests.

Providing the 'live' extraction length of the screw permits, a further advantage for flow through the width of the outlet can be gained, by slightly diverging the end walls of the hopper towards the outlet. In accordance with the principles of stress/deformation, the strength of the bulk solid between the converging walls is weakened if the minimum principal stress offered by the confining surface is relaxed. From the convention of referring to the minor principal stresses as 'sigma 2', this technique is known as 'sigma 2 relief', see Fig. 5.22.

The fact that this behaviour extends over a considerable face length of the hopper sidewalls is due to the nature of the combination of forces. A material will slide over a contact face only when the force promoting motion exceeds the frictional resistance. However, once such a force has initiated motion, the stimulus being gravity in a solids flow situation, then any additional force at a differing angle of action will combine to produce a resultant force of higher magnitude and slightly differing direction. In this case, relaxing the confinement allows the active component of the minimum principal stress to deform the cross-section of the material in flow at a smaller compressive stress than is otherwise necessary. As a consequence of sigma 2 relief, flow will take place through smaller outlets and down shallower sidewalls than would apply with parallel end walls on the hopper. In practice a few degrees declination is effective. Very diverging walls offer almost unconfined conditions, but are less practical.

Utilizing this technique requires consideration of the upper dimensions of the diverging section of the flow channel. The reverse wall angle must not unacceptably reduce the length of the flow slot, certainly not to be less

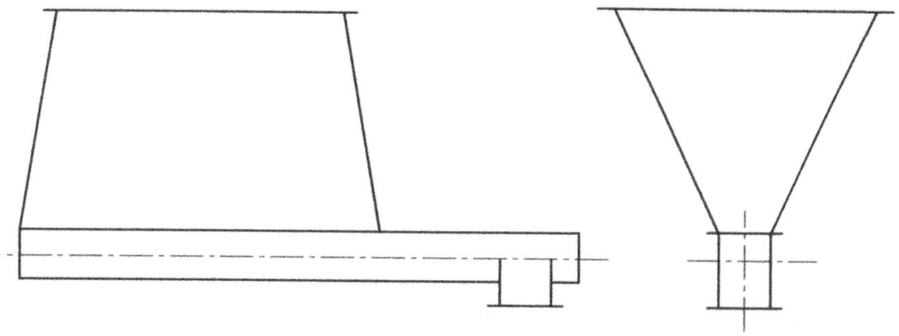

Fig. 5.22 'Sigma 2 relief' of a flow channel

than the critical 'rathole' size. There are alternative constructions for the upper section of transformation to the body of the storage container. These are to employ a non-mass flow, self-clearing design, or to extend the flow channel with a separate mass flow construction. A self-clearing section is simple to design. The decision to adopt a full mass flow design influences the lower part, as it then may be more effective to extend the diverging walls to the full width/diameter of the main body. If this is the case it will not be necessary to continue the negative rake of the sidewalls over the full distance. At a location where the span across the converging walls is well in excess of any danger of arching or ratholing, then the end walls can be made vertical, or even converging. Once the sidewall span reaches the size of the main body, then the end wall design is of similar 'vee' shape, but at 90 degrees to the lower section. Optimization of this form of construction is perhaps best left to experts, but the potential gain for flow reliability is a great attraction when dealing with difficult flow materials.

5.3.1 Inclined outlets

The design of a connection from an inclined slot outlet to the main body section of a storage container offers an interesting challenge to the designer. The conventional approach is to utilize a vertical slot of triangular elevation, to bring the connection edge to the inclined walls on to a horizontal plane, see Fig. 5.23.

This is wasteful of headroom; a more efficient way is to crank the side walls to facilitate the connection between tapering edges with a flat surface, the faces having to diverge to the same width in differing heights.

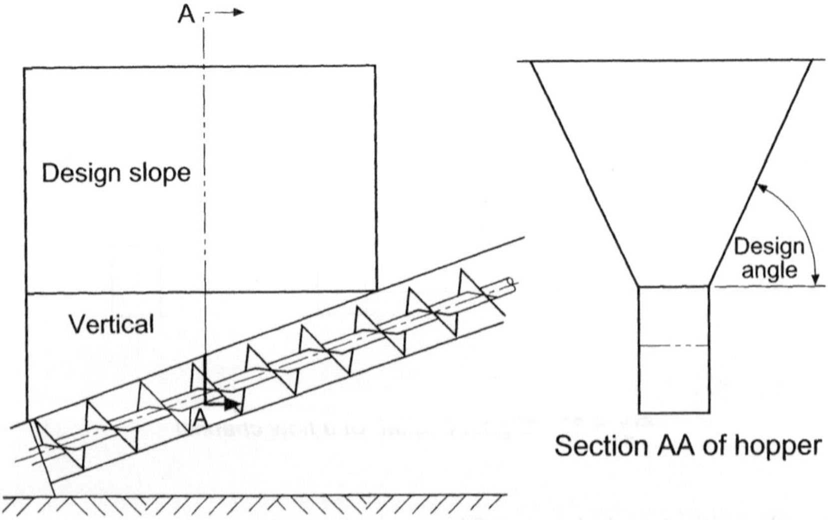

Fig. 5.23 Conventional hopper construction for inclined screw feeder

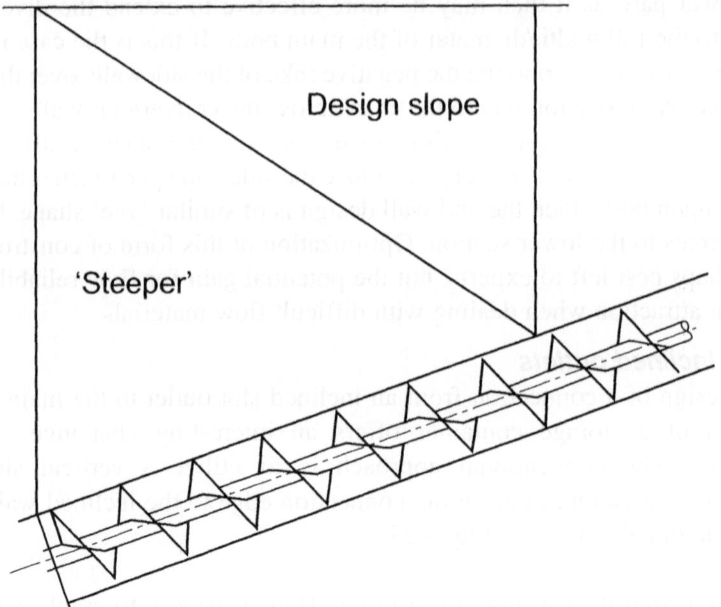

Fig. 5.24a 'Crank-in' hopper

Clearly, the headroom available at the deeper end allows the connecting edge to be steeper than the opposite end, where less depth is required to attain the same horizontal displacement. The faces of the wall formed by the cranking of the plate follow these inclinations. The crank, or 'set', in the side wall faces can be made in one of two ways, either 'crank-in' or 'crank-out', see Fig. 5.24a and b.

It will be seen that the crank-out method offers more holding capacity, but the shallow face part extends over the full length of the narrow outlet slot. By contrast, the 'crank-in' method holds less, but the steep wall part extends over the full length of the slot. The only point where the shallow face reaches the outlet is at the upper end of the outlet slot. Whether the hopper section is designed for mass flow or for self-clearing, the wall angle dictated must apply to the shallowest face in the construction. It is therefore most economic to employ the 'crank-out' form to secure the greatest spread of width, provided that the lowest wall inclination meets the criteria for flow. However, the 'crank-in' design can be used to give a form of expanded flow, with the lower section providing mass flow and the upper region being non-mass flow, mostly at a wider span of slot.

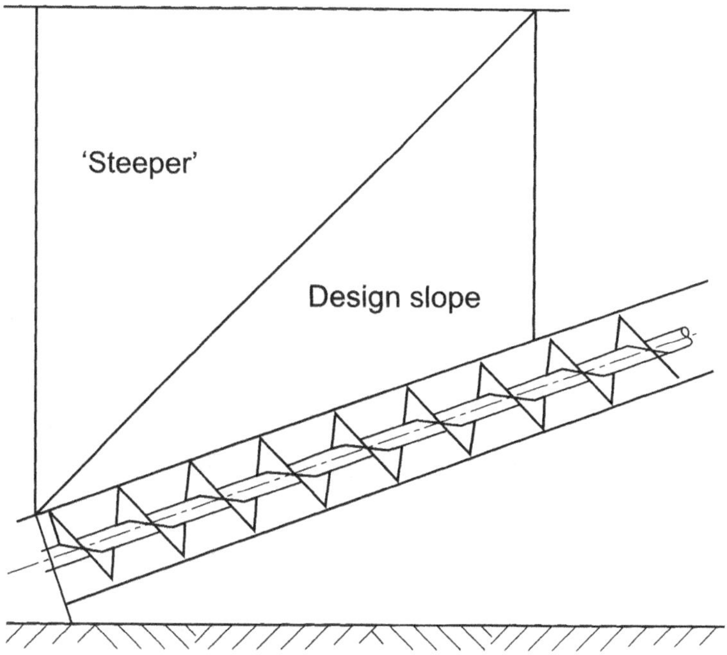

Fig. 5.24b 'Crank-out' hopper

A more efficient construction altogether is to flare the full side walls at the design angle, and make the width connection at an inclined level. There are many variations on this theme for hopper sections that do not have a large, if any, body section, as shown in Fig. 5.25a. The flare can extend an equal length, to give a low front-raised back construction for front loading and reduced spillage. The sides can be of equal height, forming a hopper with wide front and narrow back. The back face can also be made square with the outlet to give a negative rake on the back face, for more capacity and better flow.

The temptation to make the front face square with the outlet slot, as well, to simplify construction, should be resisted. This step would require the wall angles to be established on the basis of a cone inclined at the gully angle, rather than to suit a plane flow slot, and is self-defeating in both capacity and flow terms. The only time that this may be a good compromise, is when the overall length of the feeder, or of the interface connection to the hopper, would be otherwise excessive, and may be shortened by squaring off the lower face of the feeder wall to the feeder casing.

5.4 Screw extraction patterns

Very simply, the extraction pattern generated by a fully covered feeder screw is almost totally determined by the incremental transfer capacity along the screw axis caused by a change in the screw geometry. The transfer of shear forces between adjacent layers of a flowing stream tends to modify slightly the entrainment pattern by enhancing the pressure on slow-moving regions and restraining the faster flowing regions. There also tends to be a bias of flow into the 'rising' side of the rotating screw, as the angle of the flight uncovers a gravitational void. The width of the flight tip clearance and the angle of the casing wall together determine whether wall slip takes place to give inflow over the full width of the feeder casing. An essential prerequisite of a mass flow pattern in the storage container is that the total inlet area of the feeder casing area is subject to 'live' flow. Both the screw form and the casing projecting to the hopper interface, need to be carefully designed to ensure that the entire material moves, although not necessarily at a uniform pace.

For applications that do not require a mass flow pattern the only condition is that the contents will eventually self-clear to be moved away by the screw. This less demanding type of flow pattern allows flared feeder casings to be designed on 'drained repose' angles of the product.

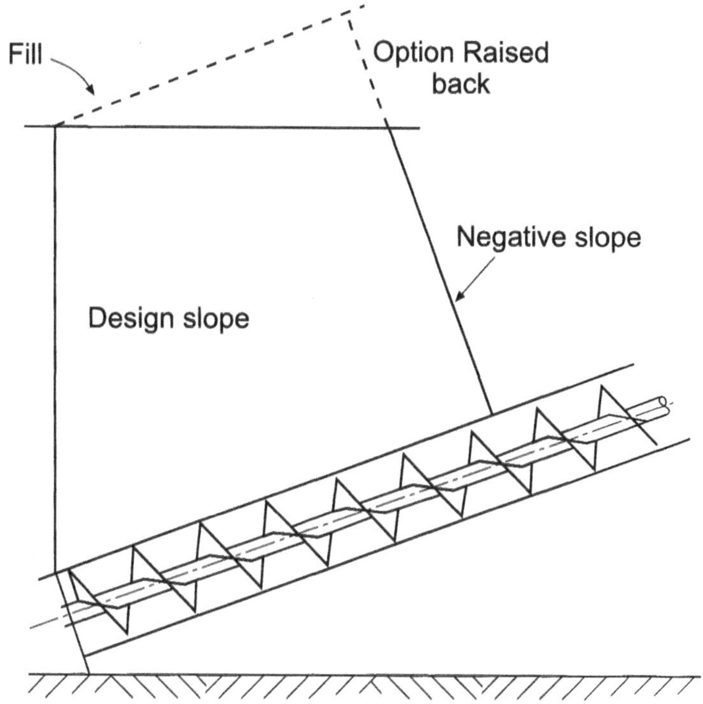

(a) 'Flared' hopper

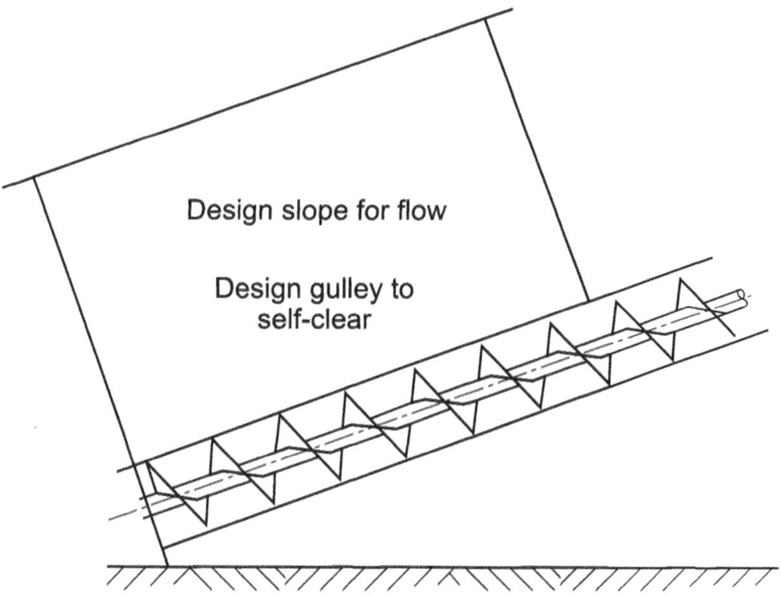

(b) Inclined standard hopper (bad design)

Fig. 5.25 Contrasting hopper forms for inclined screw feeders

There are situations which favour the use of screws with localized extraction patterns, for example to impose a degree of mixing or countering segregation in horizontal or inclined strata. In use, these screws take material from repose surface layers that cut across laminations of deposited layers, as with 'Christmas tree' segregation. They also counter the effect of mass flow velocity contours which tend to 'fold over' the boundary periphery of a deposited repose cone as the central portion is drawn down at a faster rate than the outer regions. The sophisticated use of screw extraction patterns for mixing and anti-segregation purposes requires some experience, and may call for experimental work to verify the behaviour of the bulk material in specific container geometries. The position may be summarized by three key points.

1. Local extraction is given by discrete changes of screw and shaft geometry, supplemented in some cases by the use of flow inserts to shield various portions of the screw. This form is widely used for applications that do not require a mass flow pattern, but their use relieves the power needed to drive the screw and also expands the flow channels, to avoid massive regions of static product.

2. Continuous extraction is essential for mass flow. Uneven extraction rates along the screw axis are the norm, because it is virtually impossible to secure precisely uniform extraction. The only danger is that preferential extraction will result in certain parts of the stored contents being discharged before others. This would leave the 'end of run' contents exposed to a screw form that discharges a reduced rate of feed.

3. Uniform extraction is only possible by special design over short exposed lengths.

Chapter 6

Selection Criteria

6.1 Forms of equipment

6.1.1 Standard types

The general type of screw feeder required will be clear from the nature of the duty. The choice of proprietary equipment for metering duties is often influenced by the support offered by suppliers for integrated control systems, examples of pre-proved applications, or the facility to carry out full-scale representative trials with like or similar units. For duties of discharging bins, which is usually a less onerous task by virtue of scale helping to secure reliable flow, some suppliers offer standard dimensions or recommended sizes.

The choice of standard equipment offers some advantages. The designs are complete and presumably refined by experience. Costs are known and delivery will be shorter than custom-built products, possibly even ex-stock. Variants on standards, such as differing lengths, choice of drive and its position, and connection details, can often be accommodated. The main drawback of selecting standard equipment is that construction and features special to the application cannot usually be included, except as add-ons. Equipment tailored to the duty permits features to be incorporated specific to the requirements, but can be more expensive. When formulating the specification of a unit to offer features beneficial to the particular application, the first consideration is the screw construction.

6.1.2 Screw construction

In all cases of selection, the first essential step is to decide on the form of extraction pattern that the screw has to generate over the inlet region of the casing. For any application requiring a mass flow type of extraction, the screw has to be of a progressive extraction type. This type is also useful for all long inlets. Even if the flow pattern in the supply hopper does not have to be mass flow, a screw that generates a 'live' inflow over the whole exposed length has many advantages. It absorbs less power, offers more reliable feed, and avoids any possible problem of stationary material growing firmer to arch over the screw as a result of the development of strength of the interface shear plane. The simplest method is to utilize variations of pitch over the length. Care must be taken when the changes are small, to ensure that their fitting along the screw shaft does not include any section where the pitch reduces. This expensive construction of fitting continuously variable pitches on to a tapering centre shaft is only warranted for the most demanding of applications.

Where the nature of extraction is less significant, but the inlet region is long, a stepped, graduated pitch construction is relatively economical and provides any of the benefits of a fully progressive screw construction. Even a single change is useful. Apart from forming a further in-flow channel, the shorter pitch section offers a better mechanical advantage on start-up and during running, and thereby reduces demands on the drive power.

The use of solid centre shafts is mainly confined to short feeder screws, because tubes offer much better strength-to-weight ratios for long spans. End shaft sizes are chosen to accommodate torsional and shear loads, while the main stress condition for the centre tube of the screw is that of rotary bending. Screws in regular use quickly notch up the number of revolutions to bring them into figures where fatigue stress ratings are relevant, so suitable allowance must be made for weld irregularities and stress concentrators incorporated in the construction.

6.1.3 Casing

The three common forms of casing are circular, 'U' trough, and flared 'vee'. Strictly, the circular form relates only to the section of casing outside the inlet, where the underside is normally a semicircle, limiting the inlet region to being either 'U' or 'vee' form, see Fig. 6.1.

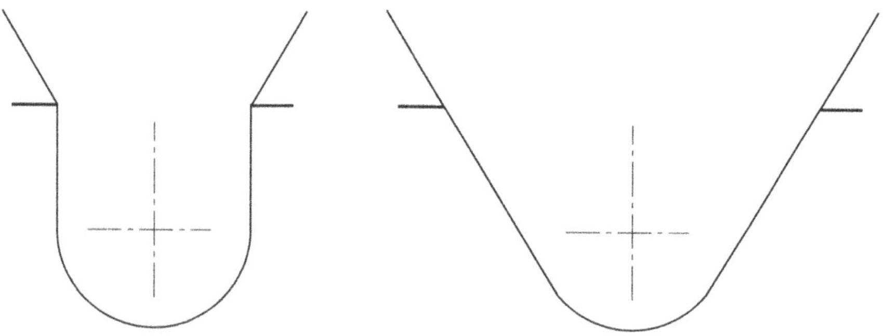

Fig. 6.1 'U' form and 'vee' form casing sections at feeder inlet

A 'vee' section of inlet diverges the casing walls to gain extra storage capacity. The inlet port can be flanged to offer a wide connection slot, or carry on as side walls of a hopper as a complete entity or to act as the base section of a larger installation. The apparent gain of effective width of opening for flow is actually a disadvantage of this construction, as the material in the clearance between the underside of the screw and the casing is static, and this layer carries round up the sloping walls to resist wall slip of the material, see Fig. 6.2. Unless the slope of the 'vee' form is exceptionally steep, this construction invariably leads to non-mass flow behaviour, whatever the form of screw construction, Fig. 6.3. A 'vee' form of casing achieves the maximum storage capacity by means of simple construction and within limited headroom, but should only be chosen when the material is of a free-flowing nature, and able to self-clear on the inclination of the wall slope formed by the casing.

A casing of 'U' form carries the clearance layer around to a vertical face. The thickness of the boundary tip clearance layer is not usually sufficiently robust to support a column of product up the wall, so that where the screw allows inflow of material over its diameter, the flow channel easily spreads to the full width of the interface with the connecting hopper. This enables a 'live' feed pattern to develop across the feeder connection. The 'U' form may also be constructed as a through casing section, or fabricated as part of a hopper which has its upper wall sections formed at an angle determined by the hopper design. Conventionally, the transition from 'U' form to hopper wall angle is made at least 25 mm above the height of the top of the feeder screw.

The section of casing projecting from the hopper length is frequently manufactured with the hopper inlet part having a 'U' or 'vee' construction.

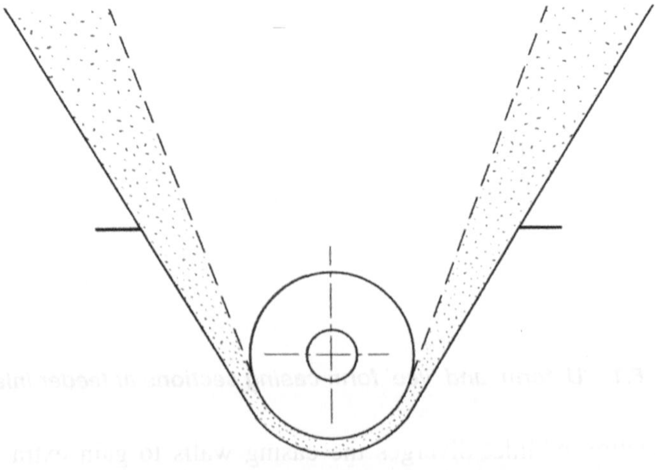

Fig. 6.2 Residue effect in 'vee' casing

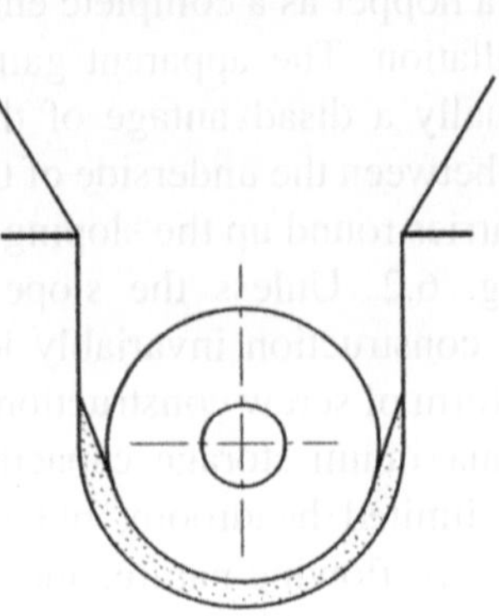

Fig. 6.3 Residue effect in 'U' casing

A choke section is then usually formed as part of the covers to restrain 'over-carry' with the screw. This form of construction allows access to inspect the screw and the outlet port by cover removal, but introduces sealing requirements not called for with extended circular sections of casing. Casing thickness is normally not a difficult issue. There is no wear by movement of material on the contact surface, unless the product is fibrous or of such a nature as to scour the clearance layer as it moves along the screw axis.

Materials of construction and finish must be selected to resist corrosion and environmental conditions. Pressures are generally low, unless there is a dust explosion danger, in which case the containment depends upon the means used to counter the hazard. Total containment requires resistance to about 8 bar, while suppression or venting is accommodated by design to hold nominally half a bar pressure. Main problem areas in such cases are bolted covers and flat surfaces, which require attention according to the relevant design for the application.

6.1.4 Covers

Covers for extended casing sections of 'U' or 'vee' form are generally flat, and folded over the sides of the casing for stiffness. Ridged forms are sometimes used to shed water on weatherproof constructions, or for stiffness on extra wide casings as used for multiple screw feeders. Hinged covers facilitate easy access, but protection for safe opening and, in the case of heavy covers, the avoidance of uncontrolled closure, should be incorporated, Fig. 6.4.

6.1.5 Seals and bearings

End seals at the discharge end of the feeder have a light duty, as the material normally falls away before reaching the point where the shaft leaves the casing. When the inlet end of the feeder is directly under the feed hopper, the local shaft seal has to cope with flooded material conditions. For crude, simple, and inexpensive applications, the end bearing is mounted directly on to the casing end plate, leaving the bearing seal to prevent leakage and bearing contamination. More usually, a gland-type seal is fitted, with space to a separately mounted pedestal bearing. This conventional approach serves most applications well, but is sensitive to shaft straightness.

Should the end shaft run out-of-true, or the gland be misaligned, the gland will suffer high radial loading as a result of the bearing alignment. Various forms of mechanical seal and machined assemblies of bearings and seals are available. In some instances, gas is injected into the seal to counter any

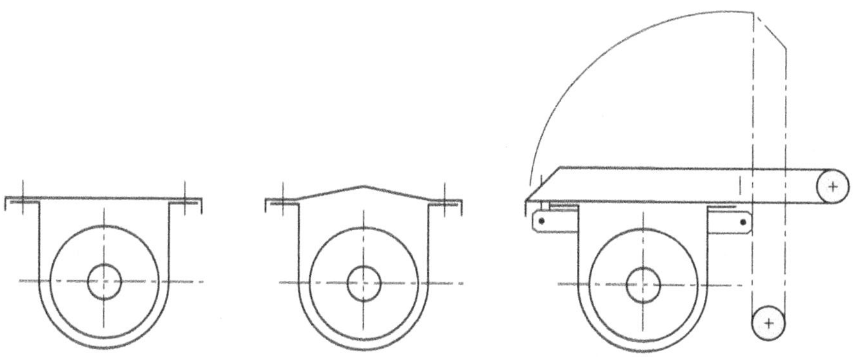

Fig. 6.4 Types of cover used on extended feeder casings

tendency for internal pressure, vapours, gases, or fine powders to leak. This mundane aspect of containment and maintenance can give rise to perennial operating problems. Provisions for seal alignment and ease of change for the packing material are useful attributes.

6.1.6 Drives

Drives for screw feeders can be fitted in the most convenient location at either end of the casing. It may be directly in line for coupling, or mounted in hollow-shaft gearboxes, above the casing at the outlet end or below the casing at the in-feed end for chain. End forces on the screw are usually accommodated by tension on the screw, therefore retention of the shaft position relative to the casing can be taken by circlips or collars on the shaft at the discharge end, irrespective of the drive location. Certification of rotation, where appropriate, should be taken from the discharge end shaft, to ensure the screw section is working. In the case of fully enclosed units, it is good practice to indicate the direction of rotation required of the drive, particularly when single or triple spur gear drive boxes reverse the motor rotation from the direction of the final screw.

Electric drives are by far the most common form of power unit. Hydraulic motors offer compactness and variability, but require a power pack and have to be sized on the maximum torque rating of the screw's duty, as they have very limited overload capacity. This feature can sometimes be used to advantage, as when delivering to a region that may block or restrict the discharge. Whereas an electrically driven feeder cannot accept a sustained stall or inhibited speed, a hydraulic drive with suitable pressure release can. A hydraulic drive can run at the set speed, at any speed restricted by the ability to clear the outlet, stop if the discharge is obstructed, and then run at the pre-set speed when the restraint is removed. Such a facility is invaluable on duties such as crammer feeders to extruders, compacting to high pressure, and for applications that are subject to erratic flow obstructions of the feeder outlet. Air motors have similar stalled torque characteristics, but some types can be load sensitive, and run slow or fast according to the resistance of the screw torque.

A host of electrical, mechanical, and hydraulic variable speed drive devices are available for screw speed control. The ubiquitous inverter offers many convenient features, not least the facility to be mounted in a safe area for feeders in hazardous zones. As with hydraulic drives, the overload capacity is small. Frame sizes are usually de-rated to compensate for lack of fan cooling at low operational speed. Auxiliary fans can be fitted if sustained, low-speed usage is expected.

6.2 Hazards and limitations

6.2.1 No feed

It is obvious that no screw feeder will work unless the material flows through the hopper outlet to fill into the volume of the screw. The hopper must therefore be designed to generate reliable flow, rather than sized to suit a particular screw sized exclusively on the feed rate to be obtained. The use of vibrators, agitators, moving walls, and other supplements to gravity, should be considered as complicating expedients to the task of securing a reliable flow of material in a consistent state of density, because their effects are not firmly predictable. The liberal use of injected air as a flow-promoting device for a feeder is particularly hazardous, because this invariably leads to uncertainty of both density and flow condition, and may lead to uncontrolled flushing.

The starting point for design is the size of orifice required for flow. It should be born in mind that the output of a feeder is controlled; therefore, it can be focused as an unconfined stream through an opening much smaller than the size of opening through which that material will pass as a confined flow channel. This feature allows the use of multiple screws to provide a large hopper outlet, with the delivery converged through a cross-section of considerably smaller dimensions. Developments in hopper design based upon measured bulk property values, and exploiting the advantages of plane flow, sigma 2 relief, and insert technology, allied to advances in feed screw design, can provide gravity flow solutions to most screw feeder applications.

6.2.2 Blocked outlet

The most common operational hazard is a blocked outlet, whether from the outlet neck of the feeder itself, or by backing up from the downstream route. Unless provision is made for tripping out the drive or limiting its power, all the reserves of torque, including whatever overload capacity is available, are concentrated on the last screw flight before the outlet, packing material in the end section of casing. Outlets can become blocked by very small restraints to the flow channel; a hand held across the outlet can be enough to hold back the discharge to start a self-sustaining blockage. Exceptional end forces can be generated without raising large pressures down the outlet port, as a consequence of the ratio of principal stresses that bulk materials can achieve and the regenerative effect of wall friction in a narrow, confined flow channel.

The usual consequence of a blocked outlet is to cause costly damage to the screw flights or to some component of the drive. This can happen quite quickly but take some time to repair, so it is good insurance to ensure this prospect is avoided. A hinged flap above the outlet fitted with a safety switch to isolate the drive provides a suitable safeguard for such eventuality. Safety clutches, shear pins, and overload protection on the drive have to allow for the normal run of occasional high loads. There is often a fine balance between building in a safety margin to avoid tripping out on occasional temporary high loads, and risking damage to the screw if the main duty is low, leaving all the surplus power available to be concentrated on the last flight. The temptation to reverse the drive in an attempt to clear a blockage must be resisted at all costs. Reversing the screw only repeats the blockage situation at the closed end of the feeder. Feed screws designed to provide a high discharge pressure need to have a robust flight construction of special design. To avoid blockages occurring as a result of sticky products adhering to the side walls, an 'expanded outlet' design directs the material to fall without contacting the discharge port surfaces. This may be constructed as a down port, or a 'through casing' construction, to offer inspection and cleaning access (Fig. 6.5).

6.2.3 Trapping in flight tip clearance

When handling fine powders the clearance between the underside of the screw flights and the casing normally fills with static product, forming a bed on which the screw moves the feed material forward. The rough boundary conditions of material composed of large particles in the moving bulk either scours the clearance space or shears a similar irregular surface. Any particle falling into this gap and becoming wedged by the moving flight, can be subjected to massive forces. These may cause the screw shaft to deflect or the drive to stall. Taper-shaped particles of a size that exceeds the clearance width, or tough flakes that can laminate, are particularly sensitive to handling by screw feeders. Large tip clearances can be used for hard irregular materials, to avoid jamming or permit the bed of product to rearrange under pressure. The clearance allowed can be varied according to the size and shape of the particles, and the particle size distribution. For

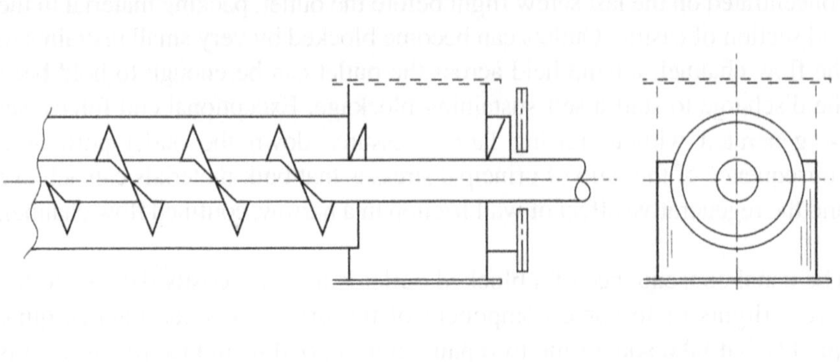

Fig. 6.5 'Expanded' outlet designs to avoid blocking

occasional large components, the probability of trapping is less than when the proportion of lumps is lower, as individual members within a mass of smaller particles will statistically tend to have infrequent contact with the underside boundary of the casing.

The prospect of particles occupying the clearance space is also dependent upon the size of screw used, larger screws being able to move larger particles within their cross-sectional area. Where there is a high proportion of larger particles in the mass, the danger of individual particles jamming, or of combinations of particles wedging in the clearance space, is increased. As a general rule of good practice, for low concentrations of large components the radial clearance should normally be at least twice the maximum particle diameter and the screw at least four times the lump size. If there is a preponderance of the larger sizes of particles, the clearance should be over four times the largest particle size and the screw over eight times the largest lump size. An alternatively approach to large tip clearances, when occasional trapping may occur, is to offset the screw axis to one side, so that the effective clearance increases as the screw rotates. This allows any trapped piece to be relieved by rotation, rather than wedging further. Products that can orient to offer an incompressible wedge, such as metal bolts and wood chips, should not be handled by screw equipment without taking special design care.

6.2.4 'Wrapping'

Long stringy products, rags, and elongated plastic, paper, rubber, or other flexible members, can wrap around the centre shaft to resist release and obstruct the passage of other materials. Shaftless screws offer good transport characteristics for such materials, provided that the surfaces are smooth and 'snag-free', and that the delivery is not at the drive end. At the drive it is necessary to employ a shaft or form of skirt to cause rotation of the screw, so components wrapped over the screw will eventually entangle at the drive end, if carried in this direction. A twin start screw is more effective in resisting such wrapping, as the items cannot wrap directly around the centre shaft but have to surround the whole flight diameter. For such applications, where practical, material should move away from the drive of a cantilever-mounted screw, so there is nothing to retain a hold on the material at the outlet.

6.2.5 'Logging' and 'clogging'

These phenomena occur when material sticks to the screw contact surface, and rotates with the screw instead of advancing along the screw axis. This behaviour tends to take place with pastes that 'shear thin' and products that have high surface cohesive or adhesive characteristics. 'Logging' is the term used when the total volume of the screw is filled and rotates. 'Clogging' takes place when material sticks to the root corners of the screw flight, usually as a large radius behind the flight face and a smaller radius on the bottom of the working face of the screw flight.

In each case the material rotates because the forces promoting the rotation exceed the forces restraining their movement. This restraint is only given by the shear strength of material in the boundary layer, or by casing friction.

Factors encouraging material to rotate with the screw are:

(a) the centre shaft, which makes no contribution to forward motion;
(b) the back face of the screw flight, which only transfers circumferential force to the product;
(c) the corner effect of the screw flight and the centre tube (which offer a large surface area in relation to the area promoting slip); and
(d) the coarse helix angle of the screw flight at the inside edge of the face – this edge has to stretch to the same pitch length as the outer tip in a much shorter circumference.

It will be seen that a flight construction that employs narrow flights, coarse pitches, rough centre tubes, irregular flight construction, weld protrusions, or high friction on flight contact surfaces, favours the rotation of product with the screw. Avoiding these features will serve most duties. The use of ribbon or 'shaftless' screws is invariably the best way to deal with poor- flow materials.

Shaftless screws have the limitation of stiffness to be self-supporting other than for short spans. Longer units run on liner plates in the casing, and may also require to be constrained from lifting in the casing. This is usually affected by fitting a longitudinal bar or angle along the casing wall on the side that the screw tends to 'climb' during rotation. Proprietary forms of ribbon flighting, having chamfered support ribs that are tangential to the centre shaft, are effective in resisting the build-up of damp products, such as filter cakes. These have the advantage of providing a support for the screw to avoid casing contact.

6.2.6 'Chatter'

A phenomenon that afflicts some extended feeder screws handling damp products is that of the screw 'chattering' on the firm residue bed of product left in the casing in the flight tip clearance. This occurs when the screw tries to ride up the side of the casing on the residue, as the rotating tip partially grips on the compacted contact surface. The influences of gravity, flight stiffness, and trough curvature to the vertical combine to limit the movement, and the screw falls back quickly to the bottom of the trough to repeat the behaviour. Serious vibration can be caused when the timing of this behaviour pattern falls near to the natural frequency of the auger shaft assembly. The phenomenon can take the form of full rotation within a circular shape of casing as a form of epicyclical rolling. there is little difference in dimensions between the effective radius of the screw and the inner surface of the residue, there is a high amplification factor to the rotational speed of the screw. One answer is to chamfer the tips of the screw flight to reduce the bearing area of the outer rim. A sharp edge tends to scrape away a layer of material and give rotational clearance for the screw to move clear of the firm bed. An alternative is to incorporate a thin blade edge-on on the rim of the screw in a coarse helical strake arrangement, oriented counter to the handing of the screw flighting. This blade scrapes away a layer of product from the clearance space in a progressive manner, allowing the screw to rotate on a fresh layer of product in transit.

6.2.7 Products that 'cake'

Damp crystalline materials that 'cake' as a mass when they dry out, products that develop hard bonds with age, and materials that compact to a firm bed under pressure, cause the material in the clearance layer to form a firm, unyielding boundary to the sweep of a feeder screw. In further use fine particles tend to trap in any minor gap formed as the flight rotates, either to build up the layer of the static bed, or to exert a binding force on the flight tip.

Such tip pressures over the working length of the screw induce a high resisting torque, due to the frictional drag on the boundary layer. High pressures and sliding resistance also exacerbate abrasive contact, to cause wear on the flight diameter. In extreme cases the pressure causes the screw to bend upwards and induce rotary bending stresses or even wear on the underside of the cover. These drag effects are magnified if the boundary layer fills all around the screw in the choke or an extended circular casing section. Not only is the area of contact increased to the full periphery of the screw, but also the screw is confined and cannot deflect to relieve the pressure. A compromise solution sometimes adopted, is to make the feeder casing of a flexible material, such as rubber belting. Periodically, when the material in the clearance layer has set hard, the casing is knocked, to deflect and fracture the thin brittle layer, and allow a further period of operation with the lining layer in a soft condition. In many cases, a thin, brittle layer is fractured by deflection under pressure from the passage of the flight tip, and the lumps are self-cleared by the conveying action.

6.2.8 *Products that 'flush'*

A screw feeder cannot restrain products in a fully fluid-like state as they flood through the casing clearances and around the screw helix without control, driven by the hydrostatic pressure of produce in the supply hopper. Even minute crevices or pinholes in the casing or gaskets will leak under these circumstances. It is the function of the supply hopper to settle any dilated incoming bulk material that behaves as a fluid, to a stable flow condition. To do this requires a period of residence commensurate with the de-aeration characteristics of the product in the given circumstances. Measurement of the product's settling characteristics and details of flow channels in hoppers, so that the product's residence time can be maximized, are discussed elsewhere in this book. Suffice to say that provision must be made to deal with initial loading, by use of a water-tight cut-off valve on the feeder, and steps taken to ensure that a critical contents level is maintained for sustained use of the feeder.

In general, all excessively dilated particulate products have no shear strength, and therefore behave in a fluid manner. Most settle rapidly to a stable condition, unless the reducing interstitial structure obstructs the escape of the ambient gas, usually air, between the particles. Settlement is driven by particle density and obstructed by fine gas escape paths, a feature of the fineness of the particles. The rate of volume decay, and the potential for a powder to raise problems by virtue of sustaining a fluidized condition, can be classified by means of Geldart's diagram, Fig. 6.6. Powders in zone C are cohesive and difficult to fluidize from a settled condition because forces between the particles resist separation by the hydrodomic force of the fluidizing gas. Unless sustained in a dilated state from a highly agitated form of preparation, pressurized gas will penetrate the bed via cracks or fissures. Powders falling into zone A have the most suitable characteristics for fluidization, having sufficient porosity to expand evenly under the action of injected gas, but having relatively low permeability so that the volume of gas required is not large and the bed is slow to settle when the gas supply is cut off. Materials classified in zone B will fluidize with a moderate gas flow, but the bed will collapse quickly to a settled state when the gas flow ceases. Coarse particles classified within zone D are very porous, hence a large amount of gas is needed to fluidize such materials and separate gas bubbles are less likely to form.

Fully settled fine particles tend to exhibit poor flow qualities; therefore, a feeder would ideally work with material in an 'easy-flow' condition, not fluid, but not fully settled. To achieve this powder state for a long-term,

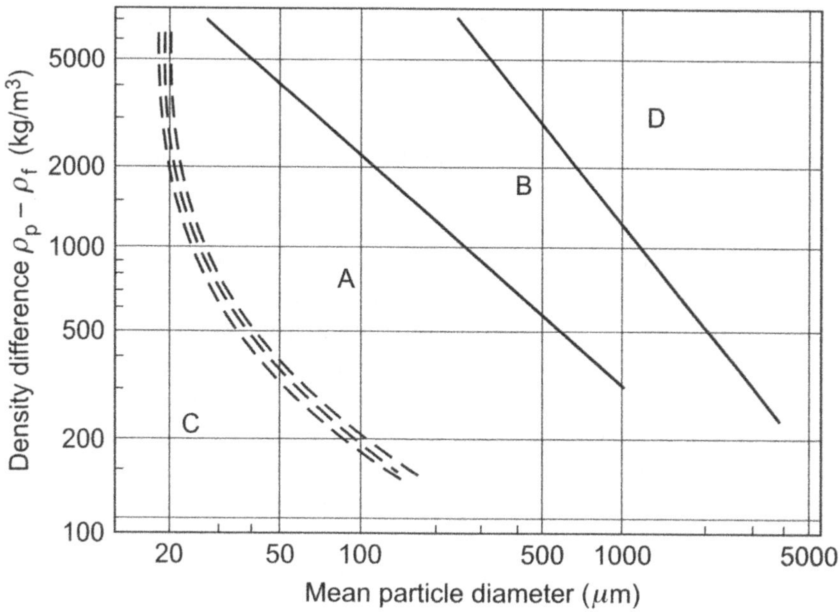

Fig. 6.6 Geldart's diagram of powder fluidization characteristics

stable operation, irrespective of whether the feeder is working or not, places high demands on a system when bulk material is received in a fluid condition. First, the incoming powder must de-aerate, to remove its excess dilation. It must then be prevented from 'loosing' further void gas at a stage where the material has developed sufficient shear strength to be stable. This must occur regardless of whether the bed is static or moving, or the feeder is operating. In practice it is virtually impossible to meet all operating conditions, as the hopper level may rise and fall according to the production circumstances, but a good compromise can usually be obtained by the use of a mass flow hopper combined with continuous, limited air injection.

Products loaded into a mass flow hopper in a fluid condition will tend to re-enforce any preferential flow channel present in the flow regime. It is therefore essential to achieve a condition of flow stability in the product before the fresh contents enter a converging region of a flow hopper, as these invariably introduce velocity variations across the cross-section. A proven system to accelerate the expression of air from a fluidized bulk material is to use a vibrating frame on to which vertical rods are mounted, which are tuned to whirl at the natural frequency of the applied vibration.

Such rods form vertical holes through which air from the deeper regions of the bed can escape to the unconfined surface of the stored material without having to pass through the tortuous path of diminishing matrix of the particle interstices.

The rate of air loss from a settling bulk mass reduces exponentially, so a small air bleed into the mass will not significantly reduce the initial settlement rate, but will increasingly resist its final settlement. Incorporating means to accelerate the de-aeration of the upper regions of storage strengthens this process. A vital feature for such a combined addition/subtraction system is that the volume of the flow channel, when the stored contents are lowest, allows this process to mature within the residence period of the product, even when the feeder is working at its maximum rate of discharge. The rate of air added depends upon many factors of the installation, but normally it is of the order of litres per hour, rather than huge gas volumes. The added air should be dry, even with non-hygroscopic bulk materials, as continuous injections of ambient air into a static bed will tend to deposit accumulated moisture.

6.2.9 Dimensional limits

Diameter Apart from the above limitations there are various size constraints to the use of screw feeders. As a result of screw boundary shear forces increasing with area exposed, i.e. as the screw diameter, and the distance at which these forces operate increase as the radius, then the torque required to start and run a screw increases at least as the square of the screw diameter. In practice, large screws serve large and long openings, so overpressures also tend to be higher to further raise these torque values. As a result, screws larger than 400 mm tend to require very heavy drives, and much larger screws than this are generally impracticable. Higher capacities and larger openings for flow are better served by the use of multiple screws than large-diameter units.

Length Under normal circumstances feed screws should not be extended by means of intermediate bearings, as these obstruct the passage of material in a filled screw and generally create objectionable effects. The maximum length of a feeder screw is therefore limited by the span that can be carried by the centre tube. This is normally determined by the amount of deflection tolerable in the casing, not torque capacity. Large tube sizes are only practical on large screws; otherwise there is not sufficient depth of flight left to be effective. The maximum length of a feeder screw is therefore related to the diameter of screw employed and the type of bearing mounting.

Cantilever-mounted screws, which allow direct end discharge, normally range in overall length from around 200 to 1000 mm for 'shaftless' screws, and from 500 to 2000 mm long for screws with centre shafts. The span of screws supported at both ends, ranges from around 2000 mm for 100 mm diameter screws, to over 6000 mm long for 400 mm diameter screws. As these lengths have to encompass the exposed length, the choke length, and the outlet port, the length of the feed hopper interface is necessarily limited, apart from considerations of providing 'live' extraction over the whole exposed length when appropriate. The opening of a fully 'live' feed channel at the hopper interface rarely exceeds a length of six times the screw diameter.

Flight sizes Good practice limits the maximum ratio of centre tube diameter to depth of the flight of feeder screws, to about 3 to 1. The flight pitch-to-depth ratio also has practical limits, to avoid 'logging' by material revolving with the screw, or excessive compacting pressures if a narrow flight width has to promote the movement of a long annulus of material. In practice the flight pitch-to-depth ratio is rarely outside the range of 0.5 to 3.

6.3 Capacity

Feed screws are essentially volumetric devices, moving material according to the swept volume and direction of travel. As outlined in Section 4.1, the path followed by material occupying the space between the flight pitches of a filled screw moves with respect to the angle of friction on the flight face, which varies in helix angle from root to tip. The amount of material carried forward within the screw volume by one rotation of the shaft is, therefore, invariably less than the swept volume by a factor generally denoted as the screw 'efficiency'.

A feature that causes most difficulty in assessing the theoretical output of a screw is that the helix angle of the flight face varies over the radius from the outside rim, where the angle is $\tan^{-1} \pi D/P$ to $\tan^{-1} \pi d/P$ adjacent to the centre shaft, where D = screw o/d, d = centre shaft diameter and P = screw pitch. A 'mean effective diameter' may be used to secure an approximate average value for the helix

$$D_m = \sqrt{\frac{D^2 + d^2}{2}}$$

This approach is based upon an equality of the areas of the flight that are 'leading' and 'lagging' the mean. The optimum pitch for maximum forward feed under this situation is when the pitch helix angle $\theta = 45° -$

$\phi_f/2$, where ϕ_f = angle of surface friction between the screw flight and the product. This leads to an optimum theoretical output per revolution for a feed screw when the pitch = πDm.tan (45°- $\phi_f/2$).

The main reason why volumetric capacity is normally estimated in a relatively crude manner, as a proportion of 'swept volume', is that derivations of transport efficiency are generally secondary to assessments of the density condition of the solid in the conveyed state. The product 'as fed' is in a dynamic condition, so the best simple assessment is to assume a 'loose poured' measurement of product density. This value can be adjusted up or down, to take account of the speed of the feed screw and the assessed packing characteristics of the flow channel from the hopper. Material carried within the normal feeder speed range of 15–100 r/min tends to be relatively dense at low speeds but more dilated at the faster speed.

There is also a tendency for the degree of screw filling to reduce at the higher rotational speeds. A gravity flow system is unable to develop sufficient force through the bulk to cause it to accelerate in both a vertical and horizontal direction and catch up to the back space of a fast-moving flight. For these reasons, the mass discharge rate falls from linearity with higher speeds of screw rotation, but within all normal operating speeds, feeder output never ceases to increase as speed increases. This is in contrast to screw elevators, where excessive speeds of rotation tend to reject material at the inlet section. These bulk dilation effects are mitigated to some extent by extended inlet sections and good shapes of in-feed channel, and accentuated by very short inlet openings to the feeder.

There is also a degree of uncertainty about the movement of product in the upper boundary clearance between the screw and the casing in the choke section. Although this clearance is small in relation to the area of the screw, any material carried along with the screw contents compensates for efficiency 'slippage'. In cases of products exhibiting high shear strength and low contact friction, as with some fibrous materials or bulk products of an overlapping nature, the flight tip clearance space under the screw may also be 'swept' by material moving along within the cross-sectional area of the screw. For small-diameter feeders, this can account for relatively large additional output, giving theoretical efficiencies of more than 100 percent. On the other hand, products that tend to stick in the corner formed between the screw flight and the centre shaft reduce the available volume to transport material. This rough surface adversely influences the degree of rotational motion of the material advanced by the screw, to further reduce the discharge rate.

It is therefore rarely warranted to attempt to calculate the efficiency of a particular screw with any precision, because of these complexities of the screw geometry and product properties. The redressing characteristic of screw feeder performance is that changing the speed of rotation of the screw can easily vary the rate of discharge. A crude estimate of feeder capacity may be obtained by assuming a general volumetric 'efficiency' of 85 percent. This may be modified by minor adjustments of a few percent, depending upon whether the flights are short or long pitch, the material has low or high friction on the flight face, and on the proportions of the centre shaft relative to the outside diameter of the screw. An alternative simple method is to use standard 'U' trough capacity tables, such as CEMA 500 and published data by many manufacturers, with the assumption that the cross-sectional loading is double the maximum 45 percent recommended for screw conveying.

Where the rate of feed relative to screw speed is sensitive it should be calibrated by trials under representative working conditions. A consequence of this lack of precision is that specifications for ranges of mass discharge should take account of a degree of safety margins. Particular care should be taken to define bounds of feeding rate in feeder specifications. Ranges should be clarified with 'inclusive', 'not less than', and 'not more than', as appropriate. It is not uncommon to find required discharge rates specified to range from zero to a specific value, implying that a variable drive can be adjusted to stable speeds from the maximum value to nought. Manufacturers of electrical variable-speed controllers claim speed-setting ranges of from 10 to 1, up to 100 to 1. In practice it is usually difficult to achieve steady running at much lower speed than given by a 10 to 1 ratio of turndown from the maximum output speed. Allowance has to be made for extra cooling at low motor speeds, and speed stability may be suspect at very low screw speeds.

Where there is a need for a wider span of speed variations, to serve a range of outputs and allow for bands of safety at each end of the range, facilities should be provided to effect a base rate speed change. This may be possible by altering chain sprockets, using pole change motors, or introducing a secondary gear or alternative gearbox to the drive arrangement. Other options include the use of twin screws with separate drives or a twin-driven epicyclic gearbox, as described in Section 7.1. Wide speed ranges should be prescribed with full appreciation and understanding of the available drive methods, both as to the speed setting and holding capacity and the effect the speed has on the available torque output.

Where the rate of feed relative to screw speed is sensitive it should be calibrated by trials under representative working conditions. The consequence of this lack of precision is that specifications for output ranges of mass discharge should include generous safety margins.

6.4 Power

The power required to drive a screw feeder is related to speed and torque taken by the screw shaft. Allowance must be made for inherent power plant losses and drag of the bearings and seals. In the case of stuffing boxes, over-tightened seals can take a surprising amount of torque. The two conditions of start-up and running torque have to be separately assessed, relative to the starting characteristics of the prime mover. Although a slow running speed absorbs more effort than higher speeds of rotation, it may be necessary to examine both ends of the range if the drive unit does not have constant torque characteristics. There is also a distinction to be made between what is expected as 'normal' running conditions, and 'exceptional' operating circumstances that may reasonably occur during the working life of the equipment. For example, starting up under load after an extended shut down may not be part of the normal duty, but could feasibly arise in the event of an emergency stop, breakdown of other plant, or following start-up after an annual holiday period or extended shut-down.

It must be emphasized that there is no relation between the power load imposed on a screw feeder and that on a screw conveyor. A feeder has to instigate shear in the inlet and promote axial movement in a confined channel, whereas a screw conveyor has only to rotate against an offset load and scrape the residue layer in unconfined conditions. The feeder duty has two components: the first relates to the flooded inlet region and the other is relevant to the subsequent transfer section. Conditions in these two sections differ for both starting and running conditions. Empty and priming conditions are ignored, as these require less power than in fully loaded conditions. For practical purposes, unless someone has zealously over-tightened a gland seal, the totally empty running torque of a screw feeder is negligible.

From the start this book is concerned with practical, rather than obtuse, theoretical, general evaluations. The subject of the power taken by a screw feeder is wide and complex, partly because of the host of variables in particulate mechanics, and partly because of the enormous range of screw types and geometry used in industry. In many instances, for small screw feeders, the range of operating variables precludes the niceties of refined calculation compared with the relative cost of differing sized power units. Abuse is widespread, and screw feeders are often key production items that are expected to suffer exceptional demands in 'normal' service.

The cost implications of large margins of uncertainty on big machines does justify a more individual approach, particularly when the application is reliability sensitive, or the size of the drive and its ramifications for cabling and controls are of an expensive nature. It is then essential to isolate what is acceptable as normal service, as 'failure' due to exceptional circumstances may be due to loading conditions well above the norm.

6.4.1 Starting conditions

In the inlet region the torque has to overcome the shear strength of the hopper interface area, that of the underside boundary area, and the resistance to frictional slip on the face of the screw flight. The shear resistance around the screw varies, because the normal force acting on the shear plane is partially active, from overpressures, and partially passive due to the confinement of the casing. The first step is to assess an average value from a knowledge or measurement of the shear strength of the material in controlled conditions. With fine powders, the shear strength under a compacting load created by a depth of bed three times the width of the hopper outlet, provides a good guide. For coarse granules, measuring the force required to shear a totally confined, settled bed, is usually enough to make the user acutely aware that provision should be made to relieve shear expansion upon starting a flooded screw in these conditions. With suitable provision the 'three times opening width' head of material will then give a usable value. This shear strength can be taken as applying radially towards the screw, all around the periphery under initial starting conditions, because the casing, rather than the screw form, mainly carries the pressure acting through the bulk.

This approach is necessarily crude, because a refined analysis has to take account of initial start compared with re-start conditions, with their totally differing stress systems in the bulk. There is also the problem that the direction of compacting stresses is not radial, so forces acting at right angles to the circumference of the screw alter throughout the periphery. With approximations being required to accommodate varying helix angles of the screw from root to tip, such simplistic steps are essential to make economic progress. Due to the inclination of the blade there is a mechanical advantage of the screw form on the failure surface. In principle, each section of a screw that has regions of differing pitches or screw geometry should be considered separately. A safe approximation is to base the design value on the longest pitch flight exposed to the hopper outlet. The calculation for torque then becomes a simple calculation as for a screw jack.

The force to be overcome, *F*, is the summation of tip shear times the boundary area of the screw. This force acts in the helical direction in which the material will move. This force translates into a normal force acting at 90 degrees to the screw blade, which in turn generates a frictional resistance to slip. Resolving these resisting forces in line and at 90 degrees to the axis of the screw, one sees, as Fig. 6.7, the axial load, *A*, and required turning force per pitch length, *T*, to be

$$A = \pi D^2 P f \cos(\phi_f + \theta_{min})$$

$$T = \pi D^2 P f \sin(\phi_f + \theta_{max})$$

where

P = screw pitch (m)

F = boundary shear strength (kg/m²)

θ = screw helix angle = $\tan^{-1}(P/\pi D)$

ϕ_f = angle of friction of the material on the face of the screw flight

The power required per pitch in kW

$$P_p = T\,(\text{Newton metres}) \times \frac{\text{r/min}}{9550}$$

In the case of variable geometry screws each pitch section should be independently calculated as their contribution to the overall values depends upon the local helix angle of the flight form. A conservative value for torque is given by basing the calculation on the angle of the largest pitch exposed in the hopper outlet and a maximum value for axial load based upon the face angle of the shortest pitch section.

$$\text{Total power kW} = \frac{T \times \text{number of flight pitches} \times \text{r/min}}{9550}$$

(due allowance being made for pitch changes)

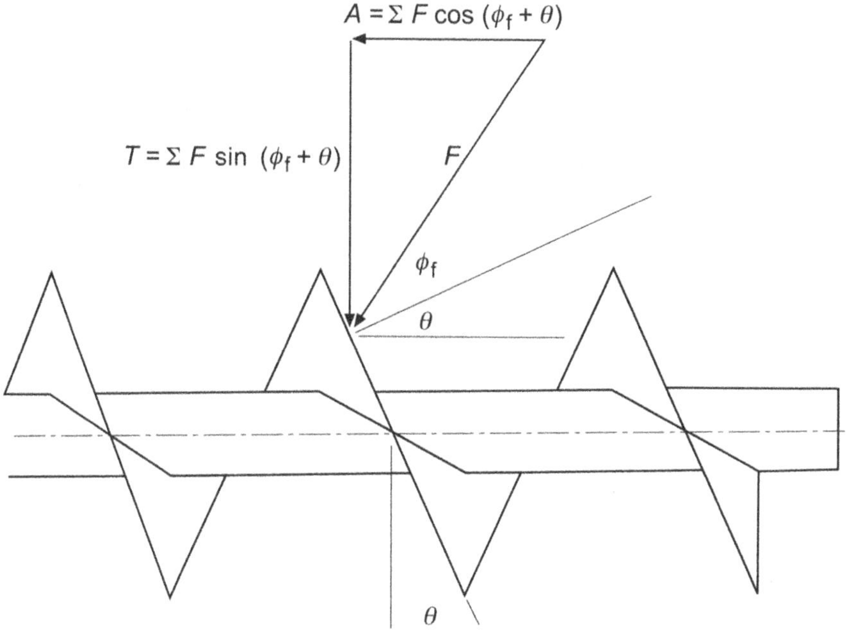

Fig. 6.7 Resolution of forces acting on screw flight

The length of screw in the hopper section is the dominant region in terms of absorbed power, and is often so with respect to its proportion of the overall screw length. For this reason, and as a safety factor, the above value may be applied to the full length of the feeder screw.

6.4.2 Running conditions

The effective geometry, dimensions, and flight face frictional value are unchanged as the screw continues to perform. The key variable is the shear strength of the interface material, which drops markedly as soon as flow commences. The reason for this fall is that material significantly dilates and shears much more easily, usually to a small fraction of its static value. The force required to shear a high flow rate stream becomes almost an insignificant factor of power requirements. At small feed rates the dilation is not as pronounced, but is still very much weaker than a settled bed. It is important to distinguish between a feeder that has a fully live flow interface with the hopper opening, and one that incorporates one or more lengths of 'dead' in-flow areas. In a 'dead' region of flow the static material will press on to the screw contents and impose a frictional drag, according to the overpressure and the internal angle of friction of the material. The underlying surface of the screw, however, still shears a confined static layer of residual product and does not diminish in the same manner. There may be some reduction in these boundary forces due to the re-orientation of particles in the shearing layer, and smoothing of the sustained shearing surface by in-filling fines.

It is normal to find that the running torque is less than half the torque required to first initiate movement, and in extreme cases can reduce to less than a tenth. This latter ratio is a reflection of the confined shear strength of overlapping granules, rather than dynamic shear values. However, as in most industrial situations the cost of retrofitting extra power is an order of magnitude greater than its cost with initial installation, it is prudent not to significantly de-rate the estimate of drive needs. Much depends upon the ratio of the starting-to-running torque characteristics of the prime mover, as to whether the drive should be sized on the starting or running torque estimates.

Allowances should be made for frequent starts, as with various types of filling and weighing operations, and with regular speed changes or sustained use at the lower end of the speed range. Special thought may be given to reducing the starting loads on feeders that are suspected to suffer high torque values at start-up. Typical means to reduce start-up loads include the injection of air or the inclusion of valves to restrict or restrain flow until the feeder is running.

Account should also be taken of potential changes of duty or use in the lifetime of the equipment. Apart from variations in the product's condition that may arise because of process variations, planned or otherwise, solids commonly vary in condition for a wide variety of reasons. Service conditions also often change in the service lifetime of the equipment. Commonly, feeder duties are up-rated for higher output, formulation changes made, variations of the product's condition, or total product changes made, without any consideration of the effect on feeder power.

A feeder that can cope with all reasonable changes of duty is a more valuable asset than one marginally able to accommodate the narrow specification of duty originally laid down for the equipment. There are many circumstances that screw feeders cannot be expected to handle, as described in the section on hazards. The effect of hard compaction of fine damp product in the working clearances is most easily overlooked, especially when caused by preferential condensation if the ambient temperature falls below dew point when handling a hygroscopic bulk product.

Chapter 7

Special Forms of Screw Feeders

7.1 Non-standard types

7.1.1 Custom built

Machines built with variations to standard construction allow special feeders to be made having casings, screws, drives, and all components constructed to sizes and incorporating features most appropriate for a specific application. Typical forms include casings and feed hoppers integrated into more complex machines, or of a geometry peculiar to a specific arrangement, feeders having special inspection access or cleaning facilities, or designs built around particular tasks and duties. An example is a feeder arrangement that serves two outlets from a common hopper, but needs to feed either or both outlets at any one time while providing a 'live' extraction from the single outlet region of the supply hopper. A single conventional reversing screw can feed either of two outlets, but not both at the same time. The need to reverse dictates that the screw construction is of uniform pitch, and to provide a 'live' extraction pattern the exposed length of screw in the hopper outlet is restricted to one-and-a-half pitches. The duty demands that the feed can be delivered to either one receiving point or the other, or both, so two feed screws are required to accommodate the latter function.

Two possible arrangements are practical. The first utilizes twin reversible screws side by side in a twin casing, having a short transverse inlet one-and-a-half pitches long by two diameters wide. Each screw is separately driven, by means of a two-speed or variable speed drive. To deliver to either side independently both screws are run at half speed to deliver the required amount. To feed both ways at once the screws are run at full speed in each direction. An advantage of this method is that, when running to one side only, the twin drives allow one screw to be stopped a short period before the other, to serve weighing and filling duties by means of a fast delivery and slow final trim output. A disadvantage is that both screws require two speed settings for one direction and another speed setting for reverse operation.

The second arrangement to meet the criteria is to have both screws end to end, running in a single casing. Cantilever-mounted screws from each end, with separate drives, then reverse at full speed to serve their nearest outlet, and run at half speed in the forward direction to feed material towards the other outlet. The hopper outlet slot is three pitches long, so half the output is taken by one screw and half by the other, whatever outlet(s) are being served (Fig. 7.1).

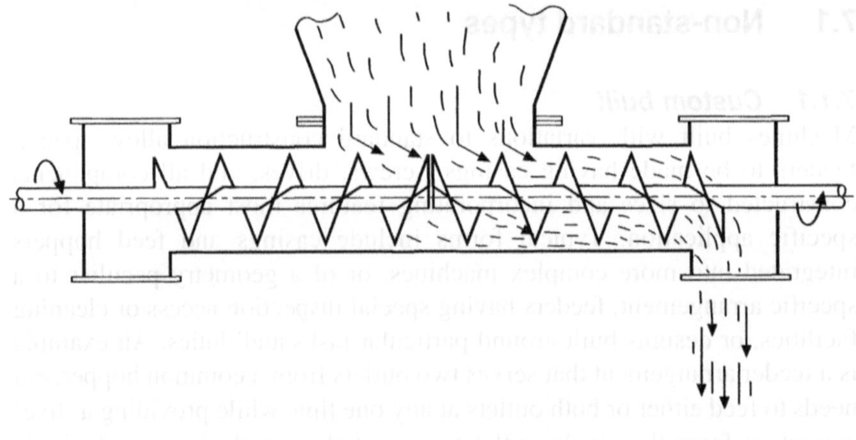

Fig. 7.1 Twin and reversing feed system

7.1.2 Hybrid feeders

Combination machines, incorporating elevation or conveying facilities, are often convenient for compactness and/or economy. Where the conveying length is such as to require intermediate bearings, then the diameter of the screw can be increased to a larger diameter in order to reduce the cross-sectional loading for the casing at the bearing location. Elevating feeders should not be inclined steeper than about 30 degrees, because it is not very practical to combine feeding at steep angles with the high-speed screw operation required to elevate at steep angles. However, stepping to an increased diameter and short-pitch flight construction, is an efficient method.

7.2 Feeders with process function

7.2.1 Pressure/heat transfer duties

Screw feeders can be designed with casing suitable for pressure or internal vacuum, or jacketed for heating or cooling the contents in transit. The heat transfer efficiency through the casing is inhibited by the presence of a layer of material in the screw flight clearance, which acts as an insulating layer between casing and product. Improved thermal efficiency is given by a heat transfer medium through the centre shaft of the screw through a rotary joint fitted to the end shaft. The options are then to use the tube surface and conduction to the screw flights, or to manufacture 'hollow' flights to carry the heat transfer surface to the wiped contact surface with the material. Whereas the 'hollow' flight method may appear the most effective, in practice it suffers from the limited interchange of product bulk with the heat transfer surface. The use of a large centre tube with a mixing flight construction to optimize the heat transfer process, provides an efficient and economic alternative.

Processes such as cooking, steaming, and sterilizing, are often enhanced by live steam injection. Similar construction allows de-gassing, inerting, and gas contact reactions to be effectively undertaken. An injection ring, with multiple entry points surrounding the screw, provides the inlet gas or steam close access to penetrate into the product.

7.2.2 'Plug seals'

A feature of some feeder operations is that the product is delivered to a region that has a slightly different pressure to that in the supply zone. Other feeders deliver into ambient conditions that are unfavourable to sustained exposure of the feeder mechanism or to the product. Common examples are feeding from dust filters, into reaction vessels that have vapour-laden atmospheres and into the high-temperature atmospheres of kilns and driers. Discharge end bearings and seals of conventional screw feeders are vulnerable to such adverse atmospheric conditions. Exposed contact surfaces suffer condensation from steamy environments, and sublimation deposits of other vapours present. Dust and fines tend to collect on exposed surfaces and accumulate to a thick encrustation when exposed to moisture. Product build-up on bearings and seal mountings, extended shafts, and on internal surfaces of casings, also causes contamination and hygiene problems, and sometimes accumulates to obstruct or totally block feeder outlets. Elevated temperatures also damage seals, bearings, and screw components. A further need when handling powders classified as a dust explosion hazard, is to provide a barrier in the flow route that will contain explosion pressures and prevent their spread from the initiating vessel.

One way to overcome these potential problems is to use a cantilever-mounted feed screw to form a 'plug seal', Fig. 7.2. Ending the screw form short of the end of the casing makes the screw push material forward to the discharge point where it falls away, to create a constantly reforming seal plug composed of the material in transit. The length of the plug is usually around one to one-and-a-half times the screw diameter, to accommodate the repose condition of the material at the end of the plug. Overlong plugs generate excessive compacting pressures on the plug, due to regenerative wall friction on the inner casing wall. Chamfering the end to the casing to match or exceed the repose angle of the product, ensures that the underside of the casing outlet is not exposed to the delivery atmosphere. It must be appreciated that such plugs need to be primed on first usage of the machine, and do not self-clear of product at the end of the run. Provision will therefore be required to clean plug seal feeders handling materials that degrade with time. Retracting drive and screw assembly types of screw feeder, as shown previously in Fig. 5.19, offer convenient access for this service.

7.2.3 Pressure seals

In some circumstances a 'plug seal' can be used as a pressure seal for delivering to a higher pressure region, such as into a reactor or a pneumatic conveying system. These applications are product dependent and technically demanding, because the permeability of the product, its internal and contact friction values, the screw geometry, and the pressure involved are all interactive. A variety of techniques are used to facilitate longer plugs and higher compacting forces. These include taper and stepped casings to relax wall pressures, and 'rifling' rods in the casing to enhance end delivery forces. In some cases weighted flaps, as with the 'Fuller'-type pump, or resisting plates are used to oppose discharge. Special screw end designs are employed to secure high compacting forces and avoid eccentric forces acting on the screw shaft.

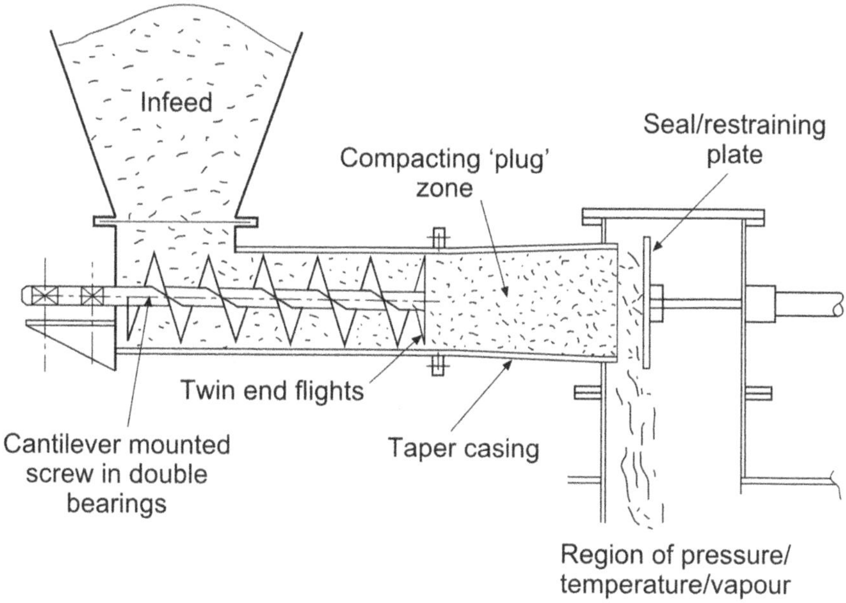

Fig. 7.2 Plug seal arrangement

7.2.4 Pressure let-down

A plug seal cannot be used on the delivery end of the feeder for pressure let-down applications, the pressure difference would simply 'blow out' the plug. In these cases, material in the infeed hopper can often be used as a pressure barrier. However, since any ambient pressure difference tends to compact the product in storage, the flow characteristics of the system have to be carefully scrutinized in the design evaluation, as does the permeability of the product. Free-flowing bulk materials tend to be of coarse composition, and therefore very permeable to gas flow. One way to avoid high gas flow rates eroding the contents of a screw feeder controlling the bulk material outflow, is to introduce a gas escape vent at the lower end of a short standpipe leading to the feeder inlet. For such applications, it is obvious that priming, low-level control, and ultimate emptying of the system are features to be specially considered.

7.2.5 De-gassing applications

In some process operations a 'de-gassing' hopper is used to change to ambient atmosphere of the void gas in a bulk material. Typical applications are the removal of volatile hydrocarbonates, toxic or noxious fumes, or the injection of a nitrogen atmosphere. Where a screw feeder is used for the discharge of such vessels, it is desirable to generate the most even flow velocity possible across the container and it is essential to avoid dead regions of flow; therefore, a mass flow form of hopper is absolutely necessary. The placement of gas injection points must also be related to the flow regime that is developed in the system, to ensure even stripping of the original void gas content of the particulate solid.

7.2.6 Injection/compaction feeders

'Plugs' can also be used with screw feeders for injecting material into another bulk mass or for compacting purposes. Confectionery examples are the use of a feed screw to inject sugar into the centre of a formed mass in the process of extrusion by a tapering bed of rotating rollers for the feed of liquorice into a multi-strand extruder. Another use of a compacting-type screw is a de-watering and extrusion machine for gluton, a maize extract, whereby the material is forced by parallel screw through a section surrounded by a 'wedgewire' screen, followed by a taper screw to enhance the delivery pressure, Fig. 7.3.

For compacting duties the nature of the bulk material bears heavily on the compacting mechanism design. Essentially there are two types: volume compaction and force compaction. Machines that compress a given volume of bulk material into a smaller volume need to be fed at a constant rate, by material in a consistent condition, so the compression ratio is constant, as determined by the change in screw geometry. Variable feed will not suit this form of machine, because a reduced amount will pass through without being sufficiently compacted. If the volume supplied changes density, the mass compacted similarly varies in final density. A particular hazard of volume compaction is that many bulk materials have highly sensitive load-compaction characteristics, i.e. they compress easily to a specific condition, but then strongly resist further reduction in volume. For example, moisture can be expressed from wet coffee grounds, but the residual cake rapidly assumes a massive resistance to further volume reduction. The critical density at which further compaction is virtually impossible is extremely sensitive, hence so are feed conditions.

For constant volume reduction the feed may be controlled by prior equipment, or the feeder served by a flooded hopper, provided the product is 'conditioned' by nature or a consistent flow channel to a consistent density. The screw design then determines the degree of compaction that the material suffers.

An alternative approach is to accept whatever rate of feed reaches the compacting screw and press this against a constant resistance to produce a controllable compacting force. This is done by a form of 'plug seal' screw, either with a fixed form of construction, or with the provision of a variable resistance device to increase or decrease the compacting force. Linking the discharge resistance to the drive unit allows the compaction force to be optimized to the drive capacity.

To densify a bulk material by expressing air from the voids, it is essential that the bulk material is not in a fluidized condition, otherwise the material will leak backwards through the screw form. When the supply material is in a highly dilated state, the feed hopper design should be of mass flow form with sufficient residence time for the material to de-aerate to a stable flow condition. Note that a mass flow hopper will not prevent the development of a preferential flow path for a bulk material in a fluidized condition. If necessary, techniques to accelerate de-aeration should be employed to secure a bulk condition that is amenable to screw compaction, e.g. vibrating rod frames, as described by the author in *Bulk Solids Handling*, 1986, Vol. 6, No. 1, p. 77.

Other examples of screw feed/process devices are feeders with pre-breaker sections, granulating, and/or sieving functions. An example of a pre-breaker with a screening section and an integral feed screw is shown in Fig. 7.4.

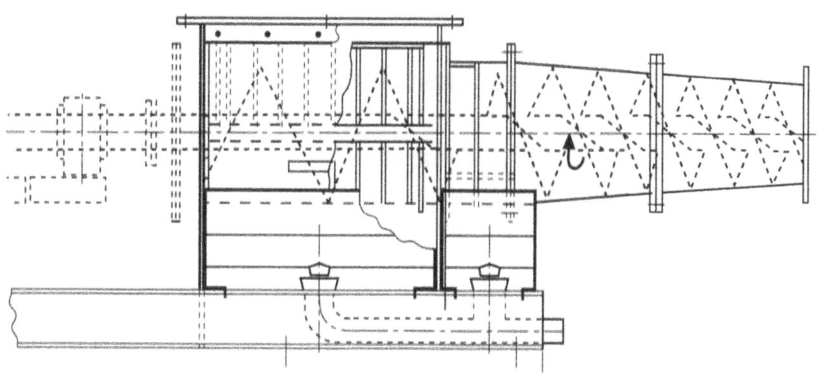

Fig. 7.3 De-watering feeder

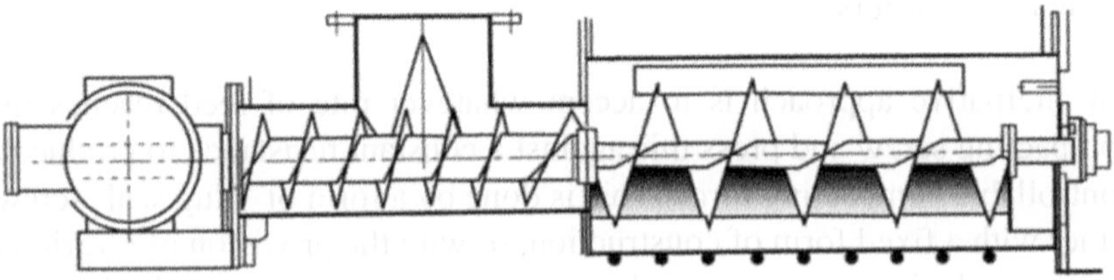

Fig. 7.4 Pre-breaking feeder

7.2.7 Blending functions

The extraction pattern of a taper screw feeder can be exploited for simple blending or homogenizing duties, by using a slot hopper with an axially adjustable wall. Moving the wall alters the proportions of ingredients taken from the two compartments. The products are mixed in transit as the screw entrains the second material and delivers the compound to the outlet. A variable plate under a shielded insert allows ingredient proportions to be finely tuned during use.

7.2.8 Crammer feeders

Screw feeders are used to deliver product into extruders, roll presses, de-aeration presses, and process machines, against a resisting pressure. They may also stimulate flow and secure feed through opening sizes much less than those over which the material will arch. Use of a horizontal screw transferring into a vertical down screw, allows a well-designed 'vee' hopper with an extended slot to provide a good flow channel, while the final screw can deliver at pressure into roll press or extruder. De-aeration tactics, as described above, may also be appropriate.

7.3 Features and accessories

7.3.1 Retracting screw assemblies

The use of cantilever-mounted screws, as shown previously in Fig. 5.19, allows the drive assembly, with machine end plates, seals, and bearings, to be withdrawn from the feeder casing without further dismantling. The drive must be isolated by means of a safety switch or interlock before the screw leaves the casing, but apart from this restriction, the method offers exceptional access for cleaning. For some applications, as with a plug seal device or compacting screw, a screw setting to adjust the distance that the screw projects into the casing provides means to vary the plug length formed, and hence the compaction condition of the product.

Short cantilever mountings can be carried directly from the drive collars of hollow-shafted geared motors. More extended lengths need to be mounted in double bearing arrangements. The limit is determined by the deflection of the screw shaft causing the flights to make contact with the feeder casing. In general, feeder lengths up to one metre present little difficulty and on units with screws above 200 mm in diameter, up to two metres' overhang can be accommodated

7.3.2 Integral end valves

The direct end outlet of cantilever-mounted screws allows the discharge port to incorporate a reverse active plate to seal the feeder when not in use. Expanding the cross-section of the outlet port and allowing the circular casing of the feeder to project slightly facilitates the mounting of an opposing air cylinder fitted with a seal plate. A feature of the arrangement is that the whole valve assembly is contained within the outlet port, hence any product falling from the feeder or valve plate discharges via the flow route rather than collecting in pockets or slide guides. The valve plate may be used in some cases, to introduce an end load to the feeder discharge, as was shown in Fig. 7.2.

7.3.3 Integral agitators

Vertical down-feed screws in a conical hopper sometimes incorporate a taper ribbon section, to feed into a small outlet neck, into which the feed screw may project. Some case is needed with such designs, because the area swept by the outer regions of the ribbon screw invariably exceeds the capacity of the central screw. With a dilated product, or one that will plastically deform, the surplus material can be expressed within the bulk, to backflow within the hopper. A firm granular product will not behave in this manner, and excessive internal stresses may result. To stimulate flow, rather than force-feed the screw, an agitator may be fitted on the vertical feeder shaft. However, a large-diameter agitator rotating at the same speed as a feed screw tends to absorb a great deal of power, and interfere with the stability of the material's density.

To overcome this problem an agitator can be mounted with an eccentric bearing on the screw shaft. Rotation of the screw does not then rotate the agitator, but causes it to 'wobble' with a significant mechanical advantage. This motion disturbs the settled condition of the material, prevents the formation of a stable arch, and stimulates flow within the hopper, so that the feed screw is filled with product in a consistent and relatively dense condition.

The field of innovation remains open to the ingenuity of the screw feeder designer. It is not possible to include in this short review more than a few examples of the variety of feeder duties and designs accomplished within the trade. The essential flexibility of helical screw equipment rests with the mechanics of the rotating wedge offered by the screw flight form. This can be seen in the wider field of applications, ranging from ships' propellers to wood screws and bolts. A prepared open mind, applied to technical challenges in the solids handling and processing industry, will find many opportunities to use screw feeders to move, meter, and condition bulk materials. It is hoped that this guide provides some useful background, stimulus, and inspiration.

Chapter 8

Case Studies

8.1 Agitated feeder

8.1.1 Background

A drier manufacturer proposed a new system to a user. The user suggested he purchased his own screw feeder to deliver the damp solid to the drier, but the manufacturer insisted the feeder was an integral part of the installation and had to be provided as part of the drying system.

Tests were conducted at the premises of a feeder manufacturer with 'representative material', to the satisfaction of all concerned, and the order progressed. A feeder costing about £2000 was incorporated in the system. Upon commissioning, it was found the feeder did not provide reliable flow due to the material 'arching'. Following some acrimonious discussion between the parties it was agreed that a second feeder with twin screws be provided at the shared expense of the feeder and drier manufacturers. However, this also proved inadequate to feed reliably, because the variable moisture content sometimes caused an intransigent flow condition of the product.

Responsibility for performance could not be agreed. The dispute culminated in litigation, the user suing for around £2 000 000 for damages, the case ultimately being settled out of court.

8.1.2 *What went wrong?*

Firstly, the specification for the material was not properly defined. The Institution of Mechanical Engineers' publication *'Guide to the specification of bulk materials for storage and handling applications'* sets out the basic handling parameters of bulk material for handling duties and highlights features of performance that should be considered.

Note It is not practical to expect any screw feeder to handle a bulk material in an undefined range of conditions. For example, what if the material freezes, contains foreign bodies, cakes, or goes off in any way? Unless the bounds of variation can be defined, the user is normally best able to assess the limits of product condition that may be presented to the feeder, but this does not mean that he accepts responsibility for performance of the equipment. In practice the user will have less understanding of the sensitive features of screw feeder behaviour. There is a need for an open exchange of information and pre-contract agreement as to the basis on which the feeder is expected to perform.

Secondly, the basis of the trials was not laid down. All the parties had different understandings of their purpose. The manufacturer expected the sample provided for test to be 'representative' and demonstrated that the equipment experienced no handling difficulty in handling this material. The users took the view that the manufacturer would interpret the test on the basis that his experience would highlight any prospective problems.

8.2 Loss in weight feeder make-up system

A loss in weight system controlling the supply of a fine powder to an extruder by a screw feeder was replenished from a make-up hopper by opening a slide valve. The make-up flow was slow to start due to an incipient arch. Then, when the arch collapsed, the material flushed through the opening to fill the feeder hopper with powder in a liquid state. The slow start of flow resulted in an unacceptably long period of non-gravimetric control. More importantly, however, the fresh material worked through the feeder hopper in preference to the residual stock, causing a total loss of feed control by flooding and overloading of the extruder.

Three things were wrong:

1. The outlet region of the make-up hopper was not properly designed

to secure reliable flow, having a small outlet and a non-mass flow shape of hopper cone.

2. The outlet size and the approach to the outlet in the make-up hopper were not designed to accommodate 'air-retarded' flow at the required discharge rate in a stable flow condition.

3. The hopper shape of the feeder and the form of the extracting screw did not *provide and sustain mass flow*, to allow fresh make-up material to stabilize in condition before reaching the discharge screw region.

Arching is caused by the outlet size being less than the 'critical arching size' for the strength of the bulk material in the particular hopper neck construction. 'Air-retarded' flow is the consequence of the limited rate of failure of the unconfined surface of the material. However, the solution required both a high flow rate and a bulk flow condition that is not excessively dilated. The 'surface area of failure' demand for this latter function would require an excessively large orifice cross-section.

One approach is 'controlled state control', which maintains a stable flow condition of the stored material in any chosen condition of density by constant but limited rate air injection. However, this technique is sensitive to many factors and usually only appropriate for relatively large storage units.

The method adopted to cure arching, 'air-retarded' flow rate, and product flow condition, was to install a 'bullet' type insert/valve in the re-supply hopper. This had the effect of disturbing the settled mass and exposing an unconfined channel as the insert was raised, so allowing flow to start easily. The confined mass flow channel formed consisted of an annular eccentric wedge shape, with an effective outlet area approximately five times the final outlet size to overcome flow rate problems. The converging neck under the insert/valve acted as a focussing chute to transfer the flowing material into the screw feeder hopper in an undilated condition.

Fortunately, because it would have been impossible to rectify the feeder hopper and screw, the resulting stable flow condition achieved was adequate for the feeder to control the outflow, despite not being of mass flow design. If the material is in a fluid state it is difficult, even for a normal mass flow hopper, to prevent a preferential flow channel developing. It is essential in such cases to provide a parallel body section to a well-designed hopper and extraction system, to give a coherent zone of 'bed flow'. Such a receiving zone allows the fresh material to settle and stabilize in its flow condition, so preventing the formation and development of a preferential route that allows material in a fluid condition to 'short circuit' the prior stock held in the storage container.

8.3 Inclined screw feeder with twin agitator

A herbicide manufacturer organized a new facility to mix, dry, and pack a new product. The engineer selected the main process and packing plant from specialized suppliers and then planned in the solids handling equipment to suit. The drier required a continuous feed from a batch mixing operation, so a fork lift/bin system was utilized to collect the mixed batch and dump the contents into a receiving hopper with screw feeder. The plant geometry and limitations on the fork truck lifting height dictated that the feeder be inclined at approximately 15 degrees to the horizontal and of extended length. Anticipating the poor flow nature of the damp powder to be handled, the engineer specified a twin agitated hopper to serve the screw feeder and secured the equipment from a local fabricator. The length of screw needed to span the feeder casing length was too great for a single span, so an intermediate bearing support was included in the design.

Although the standard of engineering construction and finish were of good quality, the performance of the unit proved to be an unmitigated disaster. The first problem was that, despite the twin agitators, the material would not flow into the screw without manual assistance, in the form of continuous poking with a long rod, and then it would only fill a short section of the screw. Inevitably, within a short period of time, damage was caused to both the agitator and the screw as the rod was frequently trapped in the mechanism. It was also virtually impossible to clear large regions of material stored in the hopper.

Material conveyed up the casing at a full cross-sectional loading blocked at the centre bearing, because of the gap in the flights and the obstruction of the bearing mounting. Reversing the screw to clear the blockage damaged the screw flighting at the in-feed end of the machine, exacerbating the feed problem. Removal of the hanger bearing in an attempt to secure material transfer allowed the screw to ride up in the casing to push the cover plates up and, because material adhered and dried on the outlet walls, the outlet of the feeder progressively blocked. The drive and screw were strengthened in order to attain a degree of production under some difficulty.

The user eventually contacted a specialized screw feeder manufacturer for an expert assessment and recommendations. A detailed report indicated that, while the concept of twin agitators over a central feed screw in a 'vee'-shaped hopper is commonly used for metering loose solids that are not free flowing, the details of the design adopted fell far short of good practice. Some of the initial operating problems and expedient repairs that had been carried out had further aggravated the problems. A number of simple independent retrofit recommendations were proposed, to facilitate virtually unattended working. Less significant improvements, more costly to implement, were given as options, and some general advice as to key features on any future installation was given.

Clearly, the fundamental error was considering that solids flow and handling in screw feeders is only a matter of providing a wide outlet in the feed hopper. As many simple screw conveyor applications are very tolerant of design variations and wide bounds of product condition, it is not uncommon for people to think that screw feeders present few technical difficulties. The devil is in the detail. Apart from the volume of the hopper being large enough to hold a full batch of product, in almost every key feature the design of the unit took no regard of the physical properties and behavioural phenomena of the bulk material. Some of the poorly designed and incorrect features built into the client's original feeder are discussed in the following subsections.

8.3.1 The screw

1. Intermediate bearing must not be fitted on to screws transferring material with a highly loaded cross-section.

2. The flighting continued too far over the outlet port. This does not allow the material to fall clear easily and tends to direct material to one side of the outlet port. Damp product discharging to a hot ambient region is likely to create vapour that condenses on internal surfaces of the equipment casing, on to which product sticks and builds up to cause blockages. Product adherence to non-contact surfaces of the screw can also accumulate to interfere with the conveying process. These circumstances can be avoided by the use of a 'plug seal', whereby the screw ends short of the outlet and pushes forward a full cross-section of material to fill the casing and seal the prior contents for the ambient condition at the delivery point. This technique is efficient, but requires careful attention to the design of the screw and the casing outlet.

There is benefit in reducing the shaft size of the screw passing over the outlet port to the smallest size practical. This construction presents the largest open area for the material to fall clear and avoid, as far as possible, the effects of material building up on the screw shaft to obstruct the outlet.

3. The flights on a long exposed inlet hopper should offer progress extraction to the contents, otherwise some regions of the contents will remain static, while freshly entered material is discharged. The flow condition is likely to deteriorate, apart from any other problem arising, from an indeterminate period of storage. A screw inclination of 15 degrees will usually clear, provided that the material in-feed is reliable. Normal proportions of screw construction do not allow effective continuous extraction to occur over inlet slots much greater than six screw diameters' axial length.

4. The screw construction had protruding radial butt welds on the working face of the flights where the individual flight sections had been joined to form a continuous helix. These welds obstructed the smooth sliding of product on the screw blade and caused local obstructions to the passage of product. There were also minor variations in the pitch spacing of the individual flights. These were welded on to the shaft in no particular order, so that as the material progressed along the screw, filling the cross-section, the product was compressed in confined conditions, giving rise to high compacting pressures and corresponding large frictional forces.

5. A flooded feed screw should never be reversed to clear a blockage, without provision for clearing the end of the screw towards which the material is conveyed. Flight damage or drive stalling is inevitable.

6. Products that pack into a hard deposit offer a high tip drag on the flight rims, can absorb excessive power, wear the flight tips, and/or bend the screw upwards as the thickness of the deposit increases. Fine, damp powders tend to bind strongly under pressure. If they also form crystal bridges to 'cake', this can have a major effect. Simple compaction tests indicate this tendency, against which consideration can be given as to the stiffness of the shaft, the flight rim width and hardness, and the screw geometry to minimize compacting forces and favour gentle forward movement of the contents.

8.3.2 The agitator shafts

1. Rotation of the agitator moved the blades well clear of the flat hopper walls and they did not sweep the full boundary length of the hopper casing. The blades comprised short flat bars, mounted square at the end of rods bolted through the shafts, which projected along the casing walls parallel to the screw and agitator shaft axis. The trailing edge of the outer face of the rotating flat bar formed a wedging surface to compact material against the hopper wall. The inclination of the damaged blades aggravated the wall pressure problem by forcing the material against the hopper wall.

2. The blades of the twin contra-rotating agitators moved downwards against the hopper walls, passing together at the centre on each rotation, tending to compact the material. Damaged blades offered mute evidence of their vulnerability to catching the rodding poles.

3. Re-synchronizing the shafts, to position them out of phase, allowed them to work with less unnecessary energy wastage. A set of new scraper blades was recommended, to sweep the full hopper walls in a systematic pattern, to be of a curved form to suit the radius of rotation, and having a chamfered leading edge. These were to be supported from the shafts by edge-on thin bars that offered low resistance to penetration of the stored mass.

8.3.3 The hopper

1. The 'vee' form of hopper, with inclined front wall oriented square to the screw axis, formed two front gullies down which the material would not flow. Material that built up on these gullies provided a steep, large-radius mass of permanent product at each side of the hopper front.

2. The flat hopper walls ran tangentially into the curve of the screw feeder casing. Apart from not being steep enough to generate wall slip, the clearance layer of product under the tip of the feeder screw provided a firm bed to resist the movement of product down the hopper walls. The agitator blade swept an internal circle in the mass, allowing a continuous arch to form under the agitator blades. An arch also formed over the blades if the hopper was filled, because of the firm support given by the low slope of the hopper walls and the layer of 'dead' material lying against them.

3. The shape of the hopper was fundamentally unsound. A convoluted shape, with twin side radii to suit the agitator and a separate semi-circular section of casing to accept the screw, is a more conventional form. This allows the agitator blade to sweep close to the whole wall surface and material resting in the tip clearance of the screw offers no obstruction to flow. The wall angles and gullies, if any, above the agitator sweep should self-clear of product. The notes in Section 5.2 on hopper design for inclined feeders outlines the options for wall shape and rim heights to suit differing methods of filling.

8.3.4 The casing outlet

1. It is good practice when handling poor-flowing products, to 'step' the width of the outlet from the screw diameter, so that the material can fall clear of the outlet side walls. Projecting the underside of the casing a short distance into the outlet also permits the material to fall from the casing without touching the underside face of the outlet. As material does not reach the furthest face, it can fall through space without any wall contact, thereby avoiding any tendency to stick.

2. In the event of outlet blockage for any reason, a port above the outlet fitted with a pressure, level, or detection switch, can be used to isolate the drive, prevent damage, and/or raise an alarm should a blockage occur. A port of this type can also be used for inspection or cleaning, utilizing the safety switch as a drive isolator when opened for access.

The costs incurred in operator and management involvement, lost production, equipment damage, plant rectification, and spoilt product, was out of all proportion to the purchase cost of the feeder, and could not be measured in financial terms alone. Supplier relationship, product quality, customer satisfaction, and operator health and safety were all involved. Fortunately, accidents were confined to equipment damage. In situations of this kind, strains on staff at all levels are adverse to morale, confidence, and smooth planning.

The moral is those contingent liabilities of poor performance should be weighed against the cost differences between equipment supplies and the relative confidence of achieving reliable design performance and expert support.

Bibliography

Addison, H. Experiments on an Archimedean screw. *Selected Engineering Papers*, No. 75 (Institution of Civil Engineers).

Amellal, K. and **Lafleur, P. G.** (1993) Computer simulation of conventional and barrier screw conveyors. *Plastics, Rubber, and Composite Processing and Applications*, **19** (4), 227–239.

Andrews, C. K. A. The performance of helical equipment for the handling of solids. ASME paper 68-MH-38.

Anonymous (1953) *Screw Conveyors and Screw Feeders*, Book 2289 (Link Belt Company).

Arnold, P. C. and **McLean, A. G.** (1980) Bulk solids storage and flow – some recent developments. Proceedings of Seminar *Powder Europa*, 22–24 January, Wiesbaden, Germany.

Arnold, P. C., McLean, A. G., and **Roberts, A. W.** (1982) *Bulk Solids Storage, Flow, and Handling,* Second Edition [University of Newcastle Research Associates (TUNRA) Limited].

Arnold, P. C. and **Roberts, A. W.** (1992) Estimation of feeder loads. Short course notes, ITC Bulk Materials Handling, Wollonwong, Australia.

Arnold, P. C. and **Roberts, A. W.** (1992) Feeding of bulk solids from bins. Short course notes, ITC Bulk Materials Handling, Wollonwong, Australia.

Australian Standard (1990) *Loads on Bulk Solids Containers*, AS 3774-1990.

Baks, A. and **Schmid, W.** (1960) *Vertical Transport Met Schroef-Transpoteurs*, Orgaan Van Het Nederlands Institute Van Register-Ingenieurs en Afgestudeerden Van Hogere Technische Scholen, Vol. 4-15e, p. 169.

Bates, L. (1969) Entrainment pattern of screw hopper discharges. *Trans. ASME, J. Engng for Industry*, **91** (2), 295–302.

Bates, L. (1986) Interfacing hoppers with screw feeders. *Bulk Solids Handling*, **6** (1),65–78.

Bates, L. (1994) The storage, feed and collection of loose solids. *Powder Handling and Processing*, **6** (2), 215–221.

Belford, A. J. R. (1965–66) *Design of an Archimedian Screw*, (DATA Publication Session).

Bell, T. (1998) Predicting the output of a screw feeder. Third World Congress on *Powder Technology*, Brighton, UK.

Bergmann, R. F. (1934) Rational means of selecting screw conveyors. *Chem. Metallurg. Engrs*, **41** (9), 470.

Bergmann, R. F. (1987) Bin design gets systems moving again. *Engineering*, **1** (2), 62–66.

British Materials Handling Board (1986) *Guidelines for Specifying Weigh Feeder Systems* (BMHB, Buckinghamshire, UK).

British Materials Handling Board (1987) Methods of measuring the physical properties of bulk powders. *BMHB Powder Testing Guide* (BMHB, Buckinghamshire, UK).

British Materials Handling Board (1998) *User Guide to Segregation* (BMHB, Buckinghamshire, UK).

Burkhardt, G. J. (1967) Effect of pitch, radial clearance, hopper exposure and head, on performance of screw feeders. *Trans. ASME, J. Engng for Industry*, 685–690.

Carleton, A., Miles, J., and **Valentin, F.** (1969) A study of factors affecting the performance of screw feeders. *Trans. ASME, J. Engng for Industry*, 329–334.

Carley, J. F. and **Struh, R. A.** Application of theory to design of screw extruders. *Industrial and Engng Chemistry*, **45** (5), 978–982.

Carson, J. (1987) Designing efficient screw feeders. *Powder and Bulk Engng*, December issue, 32–36 and 41–42.

Conner, J. H. and **Bigio, D. L.** (1993) A model for feeder–extruder reactions. *Trans. ASME*, **115** (1), 118–123.

Conveyor Equipment Manufacturers Association (1980) *Screw Conveyors*, Book 350 (CEMA).

Dec, R. T. and **Komarek, R. K.** (1994) Optimisation of roll press screw feeder design, using experimental simulation. *Proc. Powder and Bulk Solids*, May 1994, Chicago, USA, pp. 145–155.

Ephremidis, C. S. and **Henen** (1959) Schneckenforderer mit Geschloss enem Zylindrischen, Trog2 (Auger conveyors with a closed cylindrical casing), *Fordem*, **9**, p. 614.

Gutyar, E. M. (1956) *Elementarnaya Teorija Vertikalnovo Vintovovo Transportera*, (Trudy Mosk. Institute Meck. Inst I Elekt. Selsk. Khoz. im V. M).
Haaker, G., **Poppelen, M. P.**, **Jongejan, M. P.**, and **Bekhuis, J. H.** (1993) Improvement of screw feeder geometry for better draw-down performance. Proceedings of International Symposium on *Reliable Flow of Particulate Solids, II*, Oslo, Norway, pp. 551–561.

Henderson, S. M. and **Regan, W. M.** (1959) Performance characteristics of inclined screw conveyors. *Agricultural Engineering*, p. 450.

Herbsleb , G. (1982) *Werksstaffe und Korrosion*, **33**, 334.

Hudson, W. H. (1949) *Conveyors and Related Equipment*, Second Edition (Wiley, New York).

Jenike, A. W. (1976) *Storage and Flow of Solids*, Bulletin 123, (Utah Engineering Experiment Station, University of Utah, USA).

Jenike, A. W. and **Carson, J.W.** (1975) Feeding fine solids with mass flow bins. *Chem. Eng. Progr.*, **71**, 69–70.

Mallouk, R. S. and **McKelvey, J. M.** (1953) Power requirements of melt extruders. *Industrial and Engng Chemistry*, **45** (5), p. 982.

Manjunath, K. S. and **Roberts, A. W.** (19??) The mechanics of screw feeders for uniform draw-down of bulk powders from silos. *Proc. Powder Bulk Solids Conference*, Chicago, USA, pp. 171–188.

McKelvey, J. M. (1953) Experimental studies of metal extrusion. *Industrial and Engng Chemistry*, **45** (5), p. 987.

Metcalfe, J. R. (1965–1966) The mechanics of the screw feeder. *Proc. Instn Mech. Engrs*, Part 1, **180** (6).

Millier, W. F. (1958) Bucket-elevator-auger conveyors. *Agricultural Engng*, September issue, 552.

Molotov (1956) *Basic Theory of Vertical Screw Conveyors*, USSR National Institute of Agricultural Engineering, Translation 84, p. 102 (Wrest Park, Silsoe, Bedfordshire, UK).

Mubeen, F. M. (1997) Austenitic stainless steels. *Desalination and Water Refuse*, **7/3**, November/December, 20–26.

Owen, J. H. (1936) Power absorption in screw conveyors. *Engineering*, **142**, 11 September, 291.

Peart, M. (1958) Vertical augers for a silo unloader. University of Illinois Bulletin No. 631, June.

Regan, W. M. and **Henderson, S. M.** (1959) Performance characteristics of inclined screw conveyors. *Agricultural Engng*, August issue, 450.

Rehkugler, G. E. (1958) Performance characteristics of farm type auger elevators. *Stuckgut, Fordern and Heben*, **5**, p. 302 (paper presented to North Atlantic Section ASAE, 3 September 1959).

Rehkugler, G. E. (1959) Performance of auger conveyors for handling grains and ground feeds. Agricultural Engineering Extension Bulletin 325 (Department of Agricultural Engineering, New York State College of Agriculture, Cornell University, Ithaca, New York).

Roberts, A. W. (1963–1964) An investigation of grain vortex motion with relation to the performance within vertical grain augers. *Proc. Instn Mech. Engrs*, Part 1, **178** (12).

Roberts, A. W. (1991) Determining screw geometry for specified hopper draw-down performance. Proceedings of IMechE Conference *Bulk 2000*, London, UK, October, pp. 111–116.

Roberts, A. W. (1992) Determining screw feeder geometry for specified hopper draw-down performance. Workshop 7 at the Powder and Bulk Solids Conference, 11–14 May 1992, Section 4.1–15.

Roberts, A. W., Manjunath, K. S., and **McBride, W.** (1993) The mechanics of screw feeder performance for bulk solids flow control. *Trans. Mech. Engng*, **ME 18** (1), 67–73 (Institution of Engineers, Australia).

Roberts, A. W. and **Willis, A. H.** (1962) Performance of grain augers. *Proc. Instn Mech. Engrs*, **176** (8).

Schlesinger, D. and **Papkov, A.** (1997) Screw conveyor calculations based on actual material properties. *Powder Handling and Processing*, **4** (4), October/December issue, 321–325.

Stevens, G. N. (1960) Aspects of the performance of small auger grain conveyors, *J. Agricultural Engng Res.*, **11** (1), 11–18.

Van der Kooi, P. J. (1997) Silo design, well begun is half done. *Powder Handling and Processing*, **9** (4), October/December issue, 326–328.

Woodley, D. R. (1964) *Encyclopaedia of Materials Handling*, Vol. 1, pp. 336–352 (Pergamon Press, London, England).

Yu, Y. and **Arnold, P. C.** (1995) Estimate of the volumetric efficiency of a screw feeder. Proceedings of the Fifth International Conference on *Bulk Materials Storage, Handling and Transportation*, Newcastle, UK, July, pp. 517–522.

Index

Printed and bound by CPI Group (UK) Ltd, Croydon, CR0 4YY
13/09/2026
14961031-0007